Jürgen Falbe

Carbon Monoxide
in Organic Synthesis

Translated by Charles R. Adams

With 21 Figures

Springer-Verlag New York · Heidelberg · Berlin 1970

Dr. rer. nat. Jürgen Falbe, Dipl.-Chemiker

Ruhrchemie AG, Oberhausen-Holten

Translated by Dr. Charles R. Adams

Shell Development Company, Emeryville, California, USA

Revised translation of the German edition 1967

Synthesen mit Kohlenmonoxyd

(Organische Chemie in Einzeldarstellungen, Band 10)

ISBN 978-3-642-85859-8   ISBN 978-3-642-85857-4 (eBook)
DOI 10.1007/978-3-642-85857-4

Title No. 1653

Preface

This book reviews some important reactions of carbon monoxide in organic chemistry: hydroformylation, metal carbonyl- and acid catalyzed carbonylation and ring closure reactions with carbon monoxide.

It is not merely a translation of the German edition which appeared in 1967 but the text has been completely revised. This was necessary because this chemistry is rapidly developing in research as well as in technical application, which is underlined by the increase of production of e. g. oxo chemicals from about 1.4 million tons in 1967 to 2.7 million tons in 1969, nearly a doubling within 2 years.

Quite a number of new research results were published during the last two years, and these additional references have been cited in the English edition. Most of the new papers cited deal with hydroformylation reactions: however, a number of the papers reviewed also report important new aspects in carboxylation and ring closure reactions.

The author is indebted to a number of colleagues who helped to collect these new data and have given him valuable hints and would like to thank Miss I. Förster, Dr. B. Cornils, Dr. D. Hahn, Dr. P. Schneller, Dr. H. Tummes, and Dr. J. Weber for their cooperation, and to Prof. Dr. F. Piacenti (University of Pisa, Italy) for discussions on reaction mechanisms. The author is especially grateful to Dr. Charles R. Adams of the Shell Development Company, Emeryville, California, for his cooperation in translating the German text.

Jürgen Falbe

Oberhausen-Holten, June 1970

Preface to the First Edition*

Carbon monoxide chemistry has developed remarkably during the past decade due to the technical importance of the products made from it.

A number of years have passed since the last reviews have appeared [1, 10] and it thus seemed desirable to summarize the latest research and development results. The present summary was prepared on invitation of Springer-Verlag. It covers the most important syntheses with carbon monoxide in organic chemistry.

It was not the intention of the author to give a complete survey on all the possible reactions of carbon monoxide. The reactions which fall into the field of inorganic chemistry are only mentioned if necessary for the understanding of reaction mechanisms or catalyst performance. The Fischer-Tropsch synthesis [11, 12] and the reactions of ammonia and amines [13] with carbon monoxide were also omitted since they were covered extensively by other authors.

The author would like to express his thanks to a number of colleagues who made important contributions to this book. He is especially indebted to Dr. H. Kröper, Dr. N. v. Kutepow, Dr. H. J. Nienburg, Dr. W. Himmele and Dr. D. Neubauer of BASF who reviewed the chapters on the 'Roelen', 'Reppe' and 'Koch'reactions and who added valuable supplements to the manuscript of the chapters on the Reppe and Koch reaction by providing the author with previously unpublished research results.

Dr. R. F. Heck (Hercules Powder, Wilmington, Delaware, USA), Dr. L. Marko (University of Chemical Industries, Institute of Organic Chemistry and Research Group for Petrochemistry of the Hungarian Academy of Sciences, Veszprém, Hungary), and Dr. W. Kniese (BASF Ludwigshafen) helped to resolve some questions regarding certain reaction mechanisms in discussions with the author during the time he was writing the manuscript.

Valuable contributions were also made by Dr. W. Haaf of the Max Planck Institut für Kohlenforschung, Mülheim/Ruhr, and by Dr. F. Nolte,

* Organische Chemie in Einzeldarstellungen, Band 10. Berlin-Heidelberg-New York: Springer 1967.

Dr. W. Dimmling, Dr. H. Kolling, Dr. F. Schnur, and Dr. H. Tummes of Ruhrchemie AG.

Dr. J. M. J. Tetteroo (Union Kraftstoff, Wesseling) assisted the author in collecting information and data from the literature.

Miss I. Förster, Miss D. Zimmermann, A. Hack, W. Kehn and L. Mandelartz of Ruhrchemie AG assisted the author in the collection of data and in typing of the manuscript.

The author also wishes to express his thanks to all other co-workers who made contributions.

He is especially indebted to the board of Ruhrchemie AG for support of his work.

Jürgen Falbe

Oberhausen-Holten, June 1967

Contents

II. Metal Carbonyl Catalyzed Carbonylation (Reppe Reactions)

III. Carbonylation with Acid Catalysts (Koch Reaction)

IV. Ring Closures with Carbon Monoxide

Introduction

Properties and Manufacture of Carbon Monoxide

Carbon monoxide — first discovered in 1776 by Lassonne — is a colorless, odorless, very toxic, flammable gas.

The carbon monoxide molecule is electronically similar to the nitrogen molecule. However, unlike nitrogen, the molecule has a dipole moment, because the oxygen pulls electrons towards it resulting in a polarization of the molecule [803].

Carbon monoxide has a resonating structure (a, b, c)

$$:C^+ \; :\overset{..}{\underset{..}{O}}:^- \qquad :C: \; :\overset{..}{O}: \qquad :\overset{-}{C}:::O:^+$$

$$\text{(a)} \qquad\qquad \text{(b)} \qquad\qquad \text{(c)}$$

The large resonance energy of 58 kcal/mole stabilizes the molecule despite the lack of saturation of two carbon valences. For the physical properties of carbon monoxide, see the excellent review given by A. Smeeton Leah [804].

Carbon monoxide already found technical application in the 19th century in the Mond process [14]. In 1855 M. Berthelot discovered the reaction of carbon monoxide with potassium hydroxide to give potassium formate [15] and S. M. Losanitsch and H. Z. Jovitschitsch obtained formamide when they reacted carbon monoxide with ammonia. On a smaller scale carbon monoxide was used in a number of other inorganic processes.

The development of high pressure reaction vessels by the Russian artillery officer V. N. Ipatieff in 1903 and the most important and fundamental work of C. Bosch provided the tools necessary for the application of carbon monoxide in organic chemistry.

Many reactions of carbon monoxide were discovered in the following years. A number of them found industrial application in many countries.

Today carbon monoxide is readily accessible. It can be made easily by the gasification of either carbon, methane, higher boiling paraffins or crude oil. Gasification can be achieved by reacting the named materials either with air, oxygen, steam or carbon dioxide. For some operations synthesis gas (CO + H$_2$) is required which can also be made by some of the procedures mentioned.

Methods for the preparation of carbon monoxide have been reviewed by J. Schmidt [17], W. Fuchs [732] and R. V. Green [805]. A number of other recent papers provide additional data.

Commercially there are two well known processes for producing essentially pure carbon monoxide from synthesis gas, which may be generated in the Shell or Texaco process (partial combustion of methane or oil) or in the ICI process (reforming of naphtha with steam), or come from other sources such as blast furnace gas or coke oven gas. One process is based on the absorption of carbon monoxide by ammoniacal copper carbonate, acetate, or formate solutions at elevated pressure. When releasing the pressure on the copper liquor a relatively pure carbon monoxide is obtained. The other process is based on the separation of carbon monoxide from other gases such as hydrogen, nitrogen, oxygen and methane by low temperature condensation and fractionation. An excellent review on the manufacture of carbon monoxide is given by R. V. Green [805].

The reactions of carbon monoxide within organic chemistry may be classified as follows:

(1) Three-component reactions, (2) two-component reactions with carbon monoxide insertion into an existing bond, and (3) two-component reactions with ring closure.

The hydroformylation reaction which was found by O. Roelen and the carbonylation found by W. Reppe fall into the class of three-component reactions.

The hydroformylation reaction was developed by Ruhrchemie AG, Oberhausen-Holten (Germany). Teams of the firms at Ludwigshafen and Leuna of the IG Farbenindustrie have also contributed to the development of the process. Because this reaction is at present the most important technical application of carbon monoxide chemistry, this review shall start with its description since many results obtained in the course of its investigation are of importance for the reactions dealt with in the following chapters.

The results of the research work of O. Roelen and co-workers as well as the results of W. Reppe and co-workers were published in Fiat, Cios and Bios Reports [18–20] after the end of World War II before the inventors were able to complete their work and file all necessary patents. As a result, the basic technology was made freely available to everyone and was license-free at that time. This situation attracted many companies and explains the rapid expansion of these reactions worldwide.

I

The Hydroformylation Reaction
Oxo Reaction/Roelen Reaction

1. General Remarks

Hydroformylation is the reaction of an unsaturated compound (or a saturated compound which may generate an unsaturated compound) with carbon monoxide and hydrogen to yield an aldehyde. The reaction was discovered by O. Roelen in the laboratories of Ruhrchemie AG, Oberhausen-Holten in 1938 [21–23] when he tried to recycle olefins to the Fischer-Tropsch synthesis reactor. As reaction products he isolated oxygen-containing compounds which proved to be aldehydes and ketones. Roelen started extensive investigations with ethylene and obtained propionaldehyde and diethylketone as main products.

$$H_2C=CH_2 + CO/H_2 \longrightarrow H_3C-CH_2-CHO$$

$$2\,H_2C=CH_2 + CO/H_2 \longrightarrow H_3C-CH_2-\underset{\underset{O}{\|}}{C}-CH_2-CH_3$$

The reaction was named the "oxo reaction". However, in the following years it appeared that the formation of ketones occurs to a large extent only with ethylene as starting material and only to a very limited extent with other olefins. H. Adkins [24] therefore proposed to name the reaction "hydroformylation". Today both names are used.

The reaction proceeds only in the presence of catalysts. It is exothermic: the heat of reaction being 30 kcal/mole with propylene and 28–35 kcal/mole with other olefins, varying with olefin structure and molecular weight [777].

Under suitable reaction conditions the aldehydes formed may partly be hydrogenated to the corresponding alcohols by the hydrogen present. There are processes known in which this fact is used to make alcohols in a single stage. However, in most of the operating plants the aldehydes are converted into alcohols in a second stage (see chapter on technical application of the oxo reaction).

The so-called oxo alcohols made in this operation are of great industrial importance. Besides olefins a great number of other unsaturated compounds, including polymers [806, 807], may be reacted in the hydroformylation reaction.

2. Reaction Mechanism

The technical application of the hydroformylation reaction has developed rapidly due to the industrial importance of its products. Many data were obtained during this development work, which allowed H. J. Nienburg et al. [236] and A. J. M. Keulemans, A. Kwantes and Th. van Bavel [25] to lay down certain empirical rules for the oxo reaction. For quite some time these rules were the basis for the understanding of the product distribution in the hydroformylation reactions. There was no systematic investigation of the reaction mechanism of this process in the early years. Unsatisfactory analytical results were responsible for many misinterpretations. It was assumed that the hydroformylation proceeds through heterogeneous catalysis, an assumption which is supported by some authors even in the sixties [26, 27]; (as to these papers see the critical discussion in the paper of V. Macho et al. [28]).

It took quite some time before the homogeneous nature of the catalysis in the oxo reaction was clarified and secured [1, 2, 6, 29]. This was the start for further investigations on the mechanism. However, progress was made only slowly since contradictory results and opinions were published.

After the first interpretations by O. Roelen, W. Reppe et al., H. Kröper, A. R. Martin and G. Natta et al.*, it is the merit of M. Orchin et al. [30–33], I. Wender et al. [34], R. F. Heck and D. S. Breslow [35, 36, 759], L. Marko et al. [37], P. Pino et al. [89] and F. Piacenti et al. [919] to have carried out the essential experiments which allow a deeper insight into the complicated reaction sequence in which olefins are converted into aldehydes via a number of organometallic intermediates in the hydroformylation reaction.

However, even today — 32 years after the discovery of the oxo reaction — some questions remain unclarified. Some assumptions are not yet proved by experiments, others are opposed by some of the experts.

Nevertheless, it appears that the basic steps are sufficiently secured today. Summarizing the results of the authors mentioned and the results given in papers and patents of some other workers, the mechanism is very likely to be the one given by equations (1–7), demonstrated with the example of ethylene as olefin and cobalt as hydroformylation catalyst**. Calculations based on kinetic data carried out by G. P. Wesokinskij, W. J. Gankin and D. M. Rudkovskij strongly support this mechanism [906].

---

* As to the kinetics of the oxo reaction see G. Natta et al. [38, 257, 261] (comments by I. Wender, Catalysis V add to Natta's data).

** If other olefins and other metals are used, the reaction mechanism is more or less similar to the one given in equations (1–7). Since cobalt is the most common catalyst in oxo reactions all following discussions on catalysts are based on cobalt if not mentioned otherwise.

O. Roelen was the first to assume that hydrocarbonyls are the active catalysts in the hydroformylation [2, 905]. Later M. Orchin [30] and

$$2\,Co \underset{-8\,CO}{\overset{+8\,CO}{\rightleftharpoons}} Co_2(CO)_8 \underset{-H_2}{\overset{+H_2}{\rightleftharpoons}} 2\,HCo(CO)_4 \tag{1}$$

$$HCo(CO)_4 \rightleftharpoons HCo(CO)_3 + CO \tag{2}$$

$$H_2C=CH_2 + HCo(CO)_3 \rightleftharpoons \begin{array}{c} H_2C=CH_2 \\ | \\ HCo(CO)_3 \end{array} \tag{3}$$

$$\begin{array}{c} H_2C=CH_2 \\ | \\ HCo(CO)_3 \end{array} \rightleftharpoons CH_3-CH_2-Co(CO)_3 \underset{-CO}{\overset{+CO}{\rightleftharpoons}} CH_3-CH_2-Co(CO)_4 \tag{4}$$

$$CH_3-CH_2-Co(CO)_4 \rightleftharpoons CH_3-CH_2-\overset{\displaystyle O}{\overset{\|}{C}}-Co(CO)_3 \tag{5}$$

$$CH_3-CH_2-\overset{\displaystyle O}{\overset{\|}{C}}-Co(CO)_3 + CO \rightleftharpoons CH_3-CH_2-\overset{\displaystyle O}{\overset{\|}{C}}-Co(CO)_4 \tag{5a}$$

$$CH_3-CH_2-\overset{\displaystyle O}{\overset{\|}{C}}-Co(CO)_3 \quad \begin{array}{l} \overset{H_2}{\nearrow} CH_3-CH_2-CHO + HCo(CO)_3 \\ \underset{HCo(CO)_4}{\searrow} CH_3-CH_2-CHO + Co_2(CO)_7 \end{array} \tag{6}$$

$$CH_3-CH_2-\overset{\displaystyle O}{\overset{\|}{C}}-Co(CO)_4 \quad \begin{array}{l} \overset{H_2}{\nearrow} CH_3-CH_2-CHO + HCo(CO)_4 \\ \underset{HCo(CO)_4}{\searrow} CH_3-CH_2-CHO + Co_2(CO)_8 \end{array} \tag{6a}$$

$$Co_2(CO)_7 + CO \rightleftharpoons Co_2(CO)_8 \overset{H_2}{\longrightarrow} 2\,HCo(CO)_4 \tag{7}$$

M. Almasi and L. Szabo [760] showed that cobalt in various forms is converted into cobalt hydrocarbonyls under the conditions of the oxo reaction.

More recently W. J. Gankin, D. P. Krinkin and D. M. Rudkowskij [823] showed that both $Co_2(CO)_8$ and $HCo(CO)_4$ are present in the oxo reactor when cobalt is used as the catalyst — the ratio of both depending on the reaction temperature and hydrogen partial pressure. The ratio was reported to be determined by the following equation:

$$\frac{C^2\,(\text{hydrocarbonyl})}{C\,(\text{octacarbonyl})} = K \cdot p_{H_2}.$$

The temperature dependence on the equilibrium constant K is given by the following equation:

$$K = 1.365 - \frac{1900}{T}.$$

According to the above authors under conditions usually applied in technical oxo reactions about equal quantities of $Co_2(CO)_8$ and $HCo(CO)_4$ are present in the reactor.

Moreover, I. Wender, H. W. Sternberg and M. Orchin [34] showed that stoichiometric amounts of cobalt hydrocarbonyl react with olefins at atmospheric pressure and low temperatures to yield the same reaction products as are obtained under the conditions generally applied in technical operations with catalytic amounts of cobalt hydrocarbonyl [34].

Together with other observations, these results strongly support the assumption that the hydrocarbonyl rather than the dicobaltoctacarbonyl is the active catalyst.

There are two ways for the hydrocarbonyl to react with olefins to give π-complexes, which are assumed to be intermediates in the hydroformylation reaction [38, 40, 41–43].

a) Direct attack of the olefin on the central atom of the hydrocarbonyl with formation of an intermediate complex having a coordination number enlarged by one followed by elimination of one mole of CO:

$$HCo(CO)_4 \xrightarrow{\text{olefin}} HCo(CO)_4 \text{ olefin} \longrightarrow HCo(CO)_3 \text{ olefin} + CO$$
$$(S_N2 - \text{or associative mechanism})$$

b) Elimination of one mole of CO from the hydrocarbonyl with formation of an intermediate complex having a coordination number which is lowered by one compared to the hydrotetracarbonyl followed by addition of the olefin (S_N1 or dissociative mechanism); see equations (2) and (3).

R. F. Heck and D. S. Breslow favour (b) [35] and exclude (a) [830].

There are quite a number of related reactions to support this [831–834, 938, 942, 975, 986, 1035, 1041]. Thus, for comparison the reactions of cobalt hydrocarbonyl with phosphines [832–834]

$$HCo(CO)_4 + P(C_6H_5)_3 \longrightarrow HCo(CO)_3P(C_6H_5)_3 + CO$$

and of manganese hydrocarbonyl with ^{14}C labelled CO [831, 834]

$$HMn(CO)_5 + 5\,^{14}CO \longrightarrow HMn(^{14}CO)_5 + 5\,CO$$

follow a S_N1-mechanism.

Moreover the S_N1-mechanism provides a better explanation than the S_N2-mechanism for the well established decrease of the rate of reaction by increasing CO partial pressure in the hydroformylation [39, 40].

However, it has to be mentioned that the final proof for the formation of such π-complexes and the alkylcobaltcarbonyl compounds which are believed to be formed from the π-complexes according to equation (4) under the conditions of the catalytic oxo reaction is still lacking [911, 1043].

On the other hand the subsequent steps of the reaction sequence appear sufficiently secured. It could be shown that alkylcobalt tetracarbonyls are in equilibrium with acylcobalt tricarbonyls [36] and the latter with acyl-cobalt tetracarbonyls [759] (and lit. cited in ref. [759]); (see equations (5) and (5a)). Using ^{13}CO as carbonylating agent Z. Nagy-Magos, G. Bor and L. Marko [907] showed that the acyl group is formed by incorporation of a carbonyl ligand (through migration of the alkyl group to this complex bonded carbonyl ligand) whereas CO from the gas phase is taken up as a new ligand to the cobalt atom. These results are in line with earlier findings with related manganese carbonyls [64, 908–910]. Hydrogenation of the acyl compounds yields aldehydes (equation (6)). This hydrogenation can be effected by either molecular hydrogen [36] or by cobalt hydrocarbonyl [32, 33, 36, 759]. It is very likely that under the conditions of the technical oxo process (high temperature, high pressure, low catalyst concentration) the hydrogenation is mainly effected by molecular hydrogen, since the concentration of hydrocarbonyl is extremely low during the main reaction [37]. Moreover, the rate of the reaction given by equation (7), which in the case of hydrogenation by hydrocarbonyl should be an intermediate step in the reaction sequence, is slower than the hydroformylation reaction, as shown by M. Niwa and M. Yamaguchi [44]. Only at very low CO partial pressure, which generally is not used in technical operations, it may be faster than the rate of the hydroformylation (see A. Brenman et al. [45]).

However, at the end of the oxo reaction when there is no unconverted olefin left to trap the hydrocarbonyl but only acylcarbonyl, the hydrocarbonyl may to a certain extent act as hydrogenating agent. This assumption is supported by investigations of L. Marko et al. [37] who could not find any cobalt hydrocarbonyl in the crude oxo product as long as acyl-cobaltcarbonyl was present. On the other hand they found $Co_2(CO)_8$, probably formed according to (6a) and (6) + (7). But this result cannot be regarded as final proof since the hydrocarbonyl can also well be trapped by aldehyde already formed at this stage of the reaction (see chapter I.9).

Heck and Breslow assume that the formation of aldehydes proceeds via (6) rather than via (6a), since it was demonstrated in experiments at low temperature that the hydrogenation of acylcobaltcarbonyls is inhibited by a CO atmosphere [31, 35, 36]. They formulate this reduction reaction as shown in equation (6b) and (6c) [759, 857] on the top of page 8.

However, it has to be noted that the inhibition is much lower at somewhat higher temperatures [31, 35, 36]. Moreover, Marko et al. showed acylcobalt tetracarbonyls to be present in the reaction mixture of technical oxo reactions. These acylcobalt tetracarbonyls react to aldehydes in the further course of the reaction. This could proceed via the acylcobalt tricarbonyls in the equilibrium of equation (5a). However, a reaction via (6a) cannot completely be excluded from the data available so far.

$$CH_3-CH_2-\underset{\underset{O}{\|}}{C}-Co(CO)_3 + H_2 \longrightarrow CH_3-CH_2-\underset{\underset{O}{\|}}{\overset{\overset{\displaystyle H}{\mid}}{C}}-\underset{\underset{H}{\mid}}{Co}(CO)_3 \longrightarrow$$

$$\text{(6b)}$$

$$CH_3-CH_2-CHO + HCo(CO)_3$$

$$CH_3-CH_2-\underset{\underset{O}{\|}}{C}-Co(CO)_3 + HCo(CO)_4 \longrightarrow CH_3-CH_2-\underset{\underset{O}{\|}}{\overset{\overset{\displaystyle H}{\mid}}{C}}-\underset{Co(CO)_4}{\overset{\mid}{Co}}(CO)_3$$

$$\text{(6c)}$$

$$\longrightarrow CH_3-CH_2-CHO + Co_2(CO)_7$$

As indicated earlier, the hydroformylation of most olefinic starting materials yields mixtures of isomeric aldehydes (exceptions are those olefins which are symmetrical and cannot be isomerized by double bond migration such as ethylene, cyclopentene, cyclohexene etc. These olefins yield only one aldehyde).

Terminal olefins with a longer carbon chain yield straight chain aldehydes plus branched chain isomers. The ratio of the isomers formed depends on the structure of the olefin, on the catalyst used and on the reaction conditions such as temperature and hydrogen carbon monoxide, partial pressure.

Details on the influence of these parameters will be given in the following chapters. However, some fundamental facts shall be discussed in advance since they are essential for the understanding of the reaction mechanism.

Theoretically each of the C-atoms forming the double bond in the olefinic starting material may react to form a cobalt — carbon σ-bond according to equation (4). Which C-atom will be the preferred one depends on the structure of the olefin and the catalyst.

In aqueous solution cobalt hydrocarbonyl reacts like a strong mineral acid [46—51].

However, in nonpolar solvents or in the gas state it behaves like a hydride [37, 52–57]. This fact may be explained by the assumption that only part of the charge resulting from the carbon monoxide ligands is back donated to the CO-ligands in a kind of π-bond but the rest is transferred to the hydrogen resulting in a hydride character of this hydrogen. In nonpolar solvents the hydrogen from the hydrocarbonyl should add to that C-atom of a carbon double bond which carries the lowest negative charge and the remaining cobalt tetracarbonyl should add to the other C-atom. Thus, in case of reaction with a terminal straight chain olefin the hydride ion should add to C-atom no. 2 and the cobalt tetracarbonyl to C-atom no. 1, whereas in α,β-unsaturated esters such as acrylates the direction of addition should be the reverse one; hydrogen should be added to the β-C-atom and the cobalt carbonyl to the α-C-atom. This hypothesis is fully met by the experimental results.

$$H_2C{=}CH{-}R \longrightarrow H_2C{-}CH_2{-}R \xrightarrow{CO} H_2C{-}CH_2{-}R$$
$$\qquad HCo(CO)_3 \qquad\qquad Co(CO)_3 \qquad\qquad Co(CO)_4$$

$$H_2C{=}CH{-}COOR \longrightarrow H_3C{-}CH{-}COOR \xrightarrow{CO} H_3C{-}CH{-}COOR$$
$$\qquad HCo(CO)_3 \qquad\qquad\quad Co(CO)_3 \qquad\qquad\quad Co(CO)_4$$

$$R = Alkyl$$

The direction of the addition of the hydrocarbonyl to the olefin may, within certain limitations, be influenced by varying the reaction temperature and CO partial pressure or by replacing CO ligands by other complexing agents. As shown by R. F. Heck and D. S. Breslow [35, 759] and by Y. Takegami et al. [58], the olefinic compound itself may exert a strong influence. Under certain conditions the hydrocarbonyl may even be forced to react in the acid form. Thus, at low temperatures isobutylene reacts with hydrocarbonyl to form trimethylacetyl cobalt carbonyl [35, 759] nearly exclusively. The acyl compound could be trapped with triphenylphosphine. It reacted with methanol to form methyl trimethylacetate.

$$\qquad\qquad\qquad\qquad\qquad\qquad\qquad CH_3$$
$$(CH_3)_2C{=}CH_2 + HCo(CO)_4 \xrightarrow{P(C_6H_5)_3} CH_3{-}\overset{|}{\underset{|}{C}}{-}CO{-}Co(CO)_3P(C_6H_5)_3$$
$$\qquad\qquad\qquad\qquad\qquad\qquad\qquad CH_3$$

$$\xrightarrow{I_2/MeOH} (CH_3)_3C{-}COOCH_3$$

On the other hand at higher temperatures isobutylene reacts to yield methyl-3-methylbutyrate as main product with hardly any methyl trimethylacetate being formed at all.

This behavior parallels the behavior of acrylates, which react to form α-formyl compounds at low and β-formyl compounds at high reaction temperatures [35, 59–61, 759].

These findings indicate that there are two factors controlling the isomer distribution: 1) the initial direction of the addition of the hydrocarbonyl as discussed above and 2) isomerization of the resulting complexes, alkyl- or acylcarbonyls.

Heck and Breslow have postulated equilibria between the acylcobaltcarbonyls, the alkylcobaltcarbonyls and the olefin-hydrocarbonyl complexes [35] and Takegami et al. have demonstrated in quite a number of experiments that acylcarbonyls can be readily isomerized [58, 70] *. Heck and Breslow [35] explain the significant change in the isomer distribution mentioned above by the different thermal stability of the organometallic

---

* However, it must be pointed out that the results reported here have been obtained in stoichiometric experiments under very particular conditions. It is not sure whether these isomerizations occur under technical hydroformylation conditions because other reactions may be faster.

$$\begin{array}{c} \text{C--C--C} \\ \text{O=C--Co(CO)}_{3,4} \end{array} \rightleftharpoons \begin{array}{c} \text{C--C--C--C--Co(CO)}_{3,4} \\ \parallel \\ \text{O} \end{array}$$

intermediates. It is well known that trimethylacetyl cobalt tetracarbonyl is thermally less stable than n-hexanoyl cobalt tetracarbonyl or isobutyryl cobalt tetracarbonyl e. g. [35, 759]. Also 2-carbomethoxypropionyl cobalt tricarbonyl triphenylphosphine was found to be less stable than the corresponding 3-isomer [35]. Thus, it is not surprising that, e. g. at low temperatures, isobutylene reacts to trimethylacetaldehyde while at high temperatures isomerization of the organometallic intermediates leads to 3-methylbutyraldehyde.

With strongly polarized double bonds, as in conjugated unsaturated esters and in conjugated unsaturated ethers, electronic effects seem to control the direction of the addition of the hydrocarbonyl. However, with other double bonds showing only a minor polarization, steric effects may be of significant influence [62, 63, 911, 912, 956].

That such steric effects must play an important role can already be concluded from the fact that monoalkylsubstituted olefins of type (A) react faster than disubstituted ones of type (B) [199, 956].

$$R-CH\overset{\delta\ominus}{=}CH_2 \qquad\qquad \overset{R}{\underset{R}{\diagdown}}C\overset{\delta\delta\ominus}{=}CH_2$$

$$\text{(A)} \qquad\qquad\qquad \text{(B)}$$

If only electronic effects were of importance, then (B) should react faster since it is the more polarized compound.

The steric hindrance with these types of olefins can easily be demonstrated with models.

When a π-complex of type (C) reacts to the σ-complexes (D) and (E), the hindrance in (E) is larger than in (D).

$$\text{(C)} \qquad\qquad\qquad \text{(D)} \qquad\qquad\qquad \text{(E)}$$

The existence of steric hindrance is also demonstrated by the different rates of dissociation of different acylcobalt tetracarbonyls, which increase in the series acetyl < isobutyryl < trimethylacetyl from 1 to 2.1 to 86. The steric hindrance is diminished if one CO-ligand is removed [759].

The results obtained with rhodium catalysts as described in the following chapters fit into this pattern. Rhodium is considerably larger in size than cobalt, which lessens the steric crowding of the ligands surrounding the

central atom [944, 956]. Consequently larger amounts of branched products are formed with rhodium than with cobalt at the same reaction conditions. The observed higher reaction velocity which can be achieved with
rhodium carbonyls (see page 15) may also be explained by this fact.

There remains one more isomerization to be discussed. The formyl
group formed in the hydroformylation of olefins with longer carbon chains
need not necessarily be attached to one of the C-atoms having previously
formed the double bond but can also be bound to other C-atoms of the
carbon chain. This is especially the case if the addition to neither C-atom
of the double bond results in the formation of an energetically and steri-
cally favored alkylcobaltcarbonyl. As an example internal straight chain
olefins may be taken. Under favorable reaction conditions they yield nearly
the same isomer distribution as the corresponding terminal olefins do
[25]. Another example is ethyl vinylacetate which yields the same reaction
products as ethyl crotonate does.

Obviously the cobalt, once added to the olefin in the form of its hydrocarbonyl, tends to migrate to the end of the carbon chain since this is thermally and energetically the most favored isomer among the possible alkylcobaltcarbonyls [68]. It has already been mentioned that also the terminal
π-complexes are more stable than the internal ones [69]. Heck and Breslow
[35] proposed the following mechanism for this isomerization (see footnote
on page 9).

A parallel to this type of isomerization is the similar behavior of boron-carbon [71–76], silicon-carbon [77–79] and aluminum-carbon [80] compounds.

Alternatively a different mechanism was proposed by several other authors, namely the direct isomerization of the olefinic starting material to isomeric olefins which then are hydroformylated in the second stage of the reaction. However, from the excellent work of P. Pino, S. Pucci and F. Piacenti [89], who reacted optically active (+)(S)-3-methyl-1-pentene, it can clearly be seen that under normal oxo conditions little or no isomerization of the olefinic double bond occurs prior to the hydrocarbonyl addition. Such a type of isomerization is only to be expected if very low carbon monoxide partial pressures are applied [61, 88, 89, 914]. The results of Pino et al. suggest that a multistep isomerization occurs while the olefin is attached to the metal. This could be the case if the hydride transfer, which is a necessary step in this isomerization, is faster than olefin exchange. This assumption is in line with earlier work on ironcarbonyl catalyzed isomerizations.

The reasons for the well known fact that isomeric terminal and internal olefins give nearly the same distribution of isomeric aldehydes at high reaction temperature and CO-partial pressures common for technical operations or at medium temperatures and very low CO-pressure (case 1), but different distributions if low temperature and high CO-partial pressure is applied (case 2) (where higher n/iso ratios are obtained from terminal olefins than from internal ones), are still under discussion [916].

One explanation would be that in case 1 acylcobalt tricarbonyls, which readily isomerize, would be the main intermediates and that acylcobalt tetracarbonyls, which according to Takegami et al. [70, 917, 918] are much more difficult to isomerize, would be the dominating intermediates in case 2. At low CO partial pressure and high temperature olefin isomerization before hydroformylation is at least partly responsible for the nearly identical isomeric distribution of products from terminal and interdal olefins.

Mechanisms for the hydride shift which is a necessary step in the above discussed cobalt alkyl isomerization mechanism have also extensively been discussed [31, 81, 84–96]. It appears sufficiently demonstrated that both isomerization and hydrogen transport occur intramolecularly [97, 98]. A mechanism which is in line with the experimental results is the one proposed by G. L. Karapinka and M. Orchin [86] who suggest an allyl hydride shift as shown by the following formula.

$$\underset{\substack{\text{H}_2\text{C}=\text{CH}-\text{CH}_2\text{R}}}{\overset{\substack{(\text{CO})_3\\|\\\text{HCo}\\\uparrow}}{}} \longrightarrow \underset{\substack{\text{H}_2\text{C}\overset{..}{=}\text{CH}\overset{..}{-}\text{CH}-\text{R}}}{\overset{\substack{(\text{CO})_3\\|\\\text{Co}\\\text{H}\overset{\diagup\diagdown}{}\text{H}}}{}} \longrightarrow \underset{\substack{\text{CH}_3-\text{CH}=\text{CH}-\text{R}}}{\overset{\substack{(\text{CO})_3\\|\\\text{CoH}\\\uparrow}}{}}$$

An interesting result was published recently by F. Piacenti *et al.* [919] which raises new questions. Among the reaction products of the hydro-formylation of (+)-(S)-3-methyl-1-hexene they found about 3 % of opti-cally active (R)-3-ethylhexanal; this product cannot have been formed through an isomerization mechanism as outlined above since this would definitely have resulted in racemization. The authors suggest that this product was formed through direct formylation of the methyl group in position 3, as shown by the following mechansim.

$$x \rangle 1 \quad y \rangle 2 \quad z \rangle 0$$

This result shows the complexity of organometallic reactions occurring under hydroformylation conditions.

It might well be that such direct formylation of methyl groups plays an important role in formylation of the terminal C-atoms of internal olefins under reaction conditions where migration of the double bond occurs only to a small extent.

Little is known on the stereochemistry of the addition of hydrogen and the formyl group in hydroformylation. However, the results obtained in the hydroformylation of 3.4-di-O-acetyl-D-xylal [920] of 3.4.6-tri-O-acetyl-D-glucal [921] and of norbornene [1040] where deuterium was used instead of hydrogen demonstrate a cis-addition.

Contrary to many other carbonylation reactions, carbon skeleton iso-merizations are not observed in hydroformylation.

A side reaction in hydroformylation is the hydrogenation of the olefin to the saturated hydrocarbon. This is especially observed with branched or conjugated olefins [99]. For this hydrogenation Marko proposed the follow-ing mechanism.

$$RCH_2-CH_2Co(CO)_4 \rightleftharpoons RCH_2\cdots CH_2Co(CO)_3 + CO \qquad (1)$$

$$RCH_2-CH_2Co(CO)_3 + H_2 \rightleftharpoons RCH_2-CH_2CoH_2(CO)_3 \longrightarrow$$
$$RCH_2-CH_3 + HCo(CO)_3 \qquad (2)$$

Reaction (2) is competitive to reaction (5) on page 5. As to the different effects which are observed if modified metal catalysts with ligands other than CO are applied, see chapter 3.6 on catalyst modifiers.

3. Catalysts

3.1. General Remarks

The first catalyst used in the oxo-reaction was the solid Fischer-Tropsch catalyst consisting of 66% of silica, 30% of cobalt, 2% of thorium oxide and 2% of magnesium oxide. It took quite some time until the homogeneous nature of the catalysis in the oxo reaction was discovered [1, 2, 6] and proved [100]. After this discovery a large number of metals and metal salts were investigated for their use as oxo catalysts.

So far only cobalt has found technical application. However, with rhodium excellent results have been obtained in recent years. Therefore it may be assumed that rhodium will also find technical application in the near future.

3.2. Cobalt Catalysts

Cobalt may be fed to the oxo reactor as metal [819], Raney-cobalt [101, 102], hydroxide [820], oxide [820], carbonate [102, 820], sulfate [103, 104, 821], acetylacetonate, salts of fatty acids [105, 106, 822] or in form of aqueous cobalt salt solutions [107].

All the named compounds form cobalt carbonyls under the reaction conditions applied in the oxo reaction. These carbonyls are in equilibrium with cobalt hydrocarbonyl [823] which is the active catalyst in the oxo reaction [30, 108, 109, 760] see also page 4–6. Since dicobalt octacarbonyl is readily transformed into cobalt hydrocarbonyl, it is often used as a promoter in combination with other cobalt compounds.

A number of other materials were reported to act as promoters, such as zinc [110, 337], magnesium [111], aluminum [111], bismuth [112], lead [112], gold [112], mercury salts [112], palladium zeolite [113], activated carbon [45] and iron carbonyl [171, 979]. Some of these materials were reported to accelerate the hydroformylation but suppress the hydrogenation of the aldehydes formed.

If cobalt is fed to the reactor in a form other than carbonyl or hydrocarbonyl, reaction conditions must be applied which guarantee a rapid carbonyl formation. The minimum temperature for the formation of $Co_2(CO)_2$ from finely dispersed cobalt is around 50 °C at a CO pressure of 7.5 atm [117–119]. However, a rapid reaction requires at least 30–40 atm CO partial pressure and temperatures around 135–150 °C.

$$2\,Co + 8\,CO \longrightarrow Co_2(CO)_8$$

The heat of the exothermic reaction is 110 kcal per mole of $Co_2(CO)_8$ formed.

The formation of $HCo(CO)_4$ from $Co_2(CO)_8$ proceeds under relatively mild conditions [107, 120].

$$Co_2(CO)_8 + H_2 \longrightarrow 2\,HCo(CO)_4$$

It may even proceed in the absence of molecular hydrogen if compounds are present from which hydrogen can be abstracted. Such compounds may be alcohols, amines and even cyclic paraffins [121, 124]. Definite CO partial pressures are required at definite temperatures to prevent the carbonyls from decomposing (see fig. 1) [761, 808, 809].

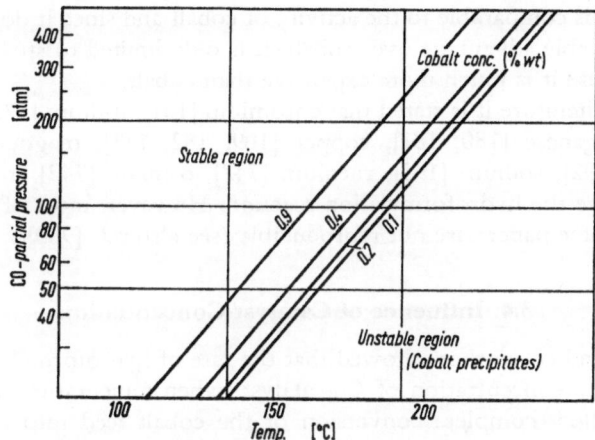

Fig. 1. Stability of cobalt carbonyl catalyst $Co_2(CO)_8 + HCo(CO)_4$ as a function of reaction temperature and carbon monoxide partial pressure in the liquid phase [761]

(Data for 0.1 and 0.2 cited from original literature, data for 0.4 and 0.9 wt % cobalt added by J. Falbe and B. Cornils (Ruhrchemie AG), based on Ruhrchemie experiments)

3.3. Other Catalysts

Iron hydrocarbonyl has been repeatedly reported to be an active hydroformylation catalyst [165, 166]. Some papers state that it is active at a lower pressure than cobalt [23, 165, 166, 168–170]. However, in a recent paper it was shown that iron hydrocarbonyl is only $1 \cdot 10^{-6}$ times as active as Co carbonyl [824, 980]. Other authors recommend adding iron carbonyl as a promoter to Co catalysts [171, 979].

Rhodium is a highly active catalyst, which was extensively investigated during recent years and has shown excellent performance [105, 106, 172–179, 811–816, 951, 956, 963, 980, 1037]. With rhodium the reaction velocity is 10^2 to 10^4 times higher than with cobalt. Thus, the catalyst concentration can be lowered to 1×10^{-3} of the concentration which is used with cobalt for the same throughput. A number of side reactions are suppressed if rhodium is applied, such as hydrogenation of the starting material and condensation and isomerization reactions. If proper conditions are applied, the hydroformylation may proceed at mild reaction conditions with high selectivity to straight chain aldehydes [817]. The kinetics of the hydroformylation with rhodium carbonyls is similar to that with cobalt hydrocarbonyl, as shown by Marko *et al.* [811, 981]. Very likely $HRh(CO)_4$ [982, 983] is the active catalyst [811, 967].

Ruthenium may also be applied as a catalyst [179, 185–190, 743]. Since its activity is comparable to the activity of cobalt and since it does not offer any remarkable advantage over cobalt, it is only limited to special applications because it is much more expensive than cobalt.

In the literature it is stated that chromium [101], iridium [173, 178, 742, 980], manganese [180, 181], copper [109, 182, 183], magnesium [191], calcium [192], sodium [193], rhenium [742], osmium [742] and platinum also catalyze the hydroformylation reaction. However, most of the results given in these papers are not reproducible (see also ref. [980]).

3.4. Influence of Catalyst Concentration

Natta and co-workers showed that the rate of hydroformylation is first order in the concentration of Co-catalyst, when temperature and CO/H_2 pressure allow complete conversion of the cobalt feed into carbonyl or hydrocarbonyl respectively [38].

This rule is invalid if cobalt is only partly converted due to insufficient carbon monoxide or hydrogen partial pressure or too low temperature. In this case the rate of formation of hydrocarbonyl is rate determining for the overall reaction [120]. In the case of low CO/H_2 pressure and high temperature, deviations may occur due to catalyst decomposition, which may cause severe problems in the oxo reactor. Cobalt deposits will reduce heat transfer to the cooling system and catalyze side reactions.

The catalyst concentration in technical operations may vary from 0.1 to 5.0 wt-% as metal, based on olefin [125], although concentrations higher than 2% are very seldom used (e. g. if low temperatures are applied in the hydroformylation). V. L. Hughes and I. Kirshenbaum reported that the catalyst concentration has a strong influence on the ratio of isomeric aldehydes formed in the hydroformylation reaction [126]. According to their publication heptene-1 will give 75% of n-alcohol plus 25% of iso-alcohol with 0.5 wt-% of cobalt applied at 85 °C and 246 atm, whereas with 2 wt-% of cobalt under the same reaction conditions only 48% of n-alcohol plus 52% of iso-alcohol should be formed.

Since no such differences in isomer distributions have ever been observed in technical operations with varying cobalt concentrations, J. Falbe, H. Tummes and J. Weber [825] have extensively investigated the isomer distribution with varying catalyst concentrations in order to check whether the reported results of Hughes and Kirshenbaum are correct. Tables 1 and 2 show the results which were obtained with propene and hexene-1 at different temperatures and catalyst concentrations.

Table 1. *Hydroformylation of propylene with different catalyst concentrations at 250 atm CO/H_2 (1 : 1)*

Temp. (°C)	Ratio cobalt to propylene (wt-%)	Ratio of straight chain to branched products	Conversion of olefin (wt-%)
90	0.5	83.5 : 16.5	73
90	1.0	83.0 : 17.0	90
90	2.5	83.1 : 16.9	90
90	5.0	83.4 : 16.6	98
90	10.0	83.2 : 16.8	95
90	15.0	83.5 : 16.5	98
120	0.2	79.2 : 20.8	85
120	0.5	79.0 : 21.0	95
120	1.0	79.0 : 21.0	96
120	2.5	78.8 : 21.2	95
120	5.0	79.0 : 21.0	95

Table 2. *Hydroformylation of hexene-1 with different concentrations of cobalt catalyst at 250 atm CO/H_2 (1 : 1)*

Temp. (°C)	Ratio cobalt to hexene-1 (wt-%)	Ratio of straight chain to branched products	Conversion of olefin (wt-%)
85	0.5	85.5 : 14.5	12
85	2.0	85.0 : 15.0	11
90	0.2	82.8 : 17.2	80
90	0.5	82.8 : 17.2	95
90	1.0	81.7 : 18.3	95
90	2.5	82.3 : 17.7	95
120	0.2	79.2 : 20.8	95
120	0.5	78.5 : 21.5	95
120	1.0	76.9 : 23.1	85
120	2.5	76.7 : 23.3	95

As can be seen from these data, the influence of the catalyst concentration on the isomer distribution of the aldehydes formed may be neglected. The difference lies within the experimental error. This is in line with results obtained with hydrocarbonyls in which one or more CO molecules were replaced by catalyst modifiers such as phosphines or phosphites [826, 827] and results which were obtained with Rh-hydrocarbonyls [828].

3.5. Catalyst Poisons

A number of compounds are known to poison the hydroformylation reaction. The poisons may act in different ways. On the one hand they may prevent hydrocarbonyl formation from the cobalt compounds fed to the oxo reactor (which would lead to troubles in discontinuous operations or in start-up in a continuous operation), on the other hand they may react with finished hydrocarbonyl to form compounds which are insoluble in the reaction mixture or inactive as oxo-catalyst.

The degree to which these compounds are of harm to the oxo-reaction may be quite different in both ways.

Thus, larger amounts of oxidizing gases such as oxygen, carbon dioxide and water [127, 128] may inhibit the formation of carbonyls from cobalt metal by passivation of the metal [119].

Fig. 2 shows the inhibition of the hydroformylation by small amounts of oxygen.

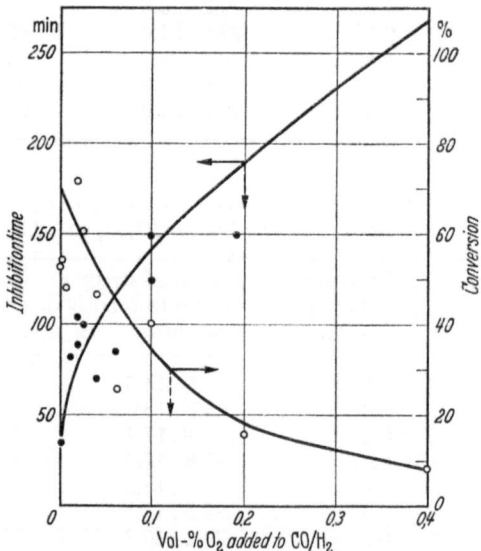

Fig. 2 [129]. Influence of oxygen on inhibition time and conversion in the hydro-formylation reaction of C_6–C_8 olefins

• Inhibition time o Conversion

It can be seen that even oxygen concentrations below 1 vol-% have a strong influence.

The same gases may also react with preformed carbonyl or hydrocarbonyl [108]. However, once the carbonyls are formed and a continuous operation is running, their influence is smaller.

Thus, with cobalt hydrocarbonyl being used as catalyst, oxygen contents up to 2 vol-% and carbon dioxide contents up to 4 vol-% in the synthesis gas feed may be tolerated, as well as water contents in the liquid feed up to 5–6 wt-% in a running reaction*.

However, with phosphine-modified catalyst systems (see chapter I/3.6), oxygen and water should be completely excluded since they would react with the phosphines to produce inactive phosphine oxides.

According to previous reviews, ammonia and amines also retard hydroformylation. This statement has to be corrected according to more recent work [7, 130, 131, 1042]. Contrary to the statement mentioned, the hydroformylation reaction is accelerated by small amounts of certain amines having an ionization constant of $\leq 10^{-8}$ [130], as e.g. pyridine, chinoline, picoline, lutidine, aniline, toluidine, xylididine, methyl aniline and aliphatic and aromatic amides such as N,N-dimethyl formamide, N-methyl pyrrolidone and acetanilide. If larger amounts of these amines are used with $HCo(CO)_4$ the effect becomes smaller. Very large amounts hinder the reaction (see table 3).

Table 3. *Influence of pyridine concentration on hydroformylation of ethylene* Solvent: toluene, 90 atm, 130 °C

Pyridine (millimoles)	$Co_2 (CO)_8$ (millimoles)	Reaction time (min) for complete reaction
0	30	36.0
30	30	9.2
60	30	5.8
120	30	7.0
250	30	8.3
1000	30	10.3
2000	30	Incomplete reaction

These results are in line with results of I. Wender *et al.* [34] who reported that a hydroformylation reaction does not take place if pyridine is used as solvent.

---

* However, there are technical processes known working even with higher water concentrations, see chapter on industrial hydroformylation operations.

Nitrogen bases having an ionization constant $> 10^{-8}$, such as N-butyl-amine, diethylamine, piperidine and triethylamine, are reported to suppress the hydroformylation completely [130].

According to V. Marko and M. Ciha [132], aqueous ammonia in concentrations of 2 to 50 wt-% on catalyst accelerate the hydroformylation reaction by accelerating the formation of cobalt carbonyl. The same explanation was given by R. Iwanaga for the acceleration by amines [133]. Ammonia concentrations higher than 50 wt-% on catalyst hinder the hydroformylation reaction (see fig. 3) [134].

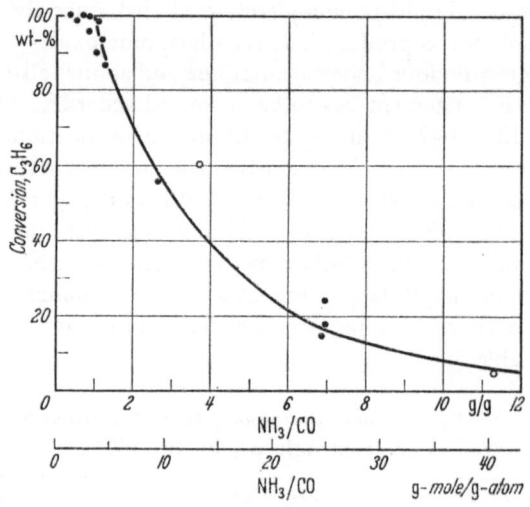

Fig. 3 [134]. Conversion of propylene in dependence on wt.- and mol-ratio NH_3/CO

● 0.20 wt % Co/C_3H_6 ○ 0.126 and 0.40 wt % Co/C_3H_6

For a long time sulfur and sulfur-containing compounds were not regarded as poisons in the hydroformylation reaction, and some authors even recommended adding small amounts of sulfur-containing compounds as accelerators.

However, more recent experiments of V. Macho [135, 136] and L. Marko et al. [137–144, 829, 913, 953–955] showed that these assumptions have to be corrected. They proved that some sulfur compounds (e. g. saturated thioethers and thiophene) are practically of no harm to the hydroformylation, but that other compounds (e. g. COS, H_2S, unsaturated thioethers, mercaptans, mercaptals, disulfides, CS_2 and elementary sulfur) inhibit the oxo reaction by forming sulfur-containing cobalt carbonyls which are inactive as oxo catalysts and less soluble than carbonyls. In the course of the reaction these carbonyls may even be converted stepwise into CoS. However, since the sulfur compounds only poison stoichiometric amounts of cobalt or carbonyls, their influence can be overcompensated by increasing the amount of cobalt in the feed and removing the CoS in the catalyst regeneration step. By this means it is possible to convert even cracked olefins containing large amounts of sulfur compounds without desulfurization prior to hydroformylation.

Nevertheless, care should be taken if high amounts of sulfur are present in the reaction mixture because sulfur may lead to trouble in the workup. Thus, sulfur compounds may be carried into the hydrogenation reactors together with the aldehydes and poison the heterogeneous hydrogenation catalysts.

Acetylene and acetylenic compounds also hinder the hydroformylation reaction [145]. The inhibition of the oxo reaction can be explained by complex formation, following the equation

$$R-C\equiv C-R' + Co_2(CO)_8 \longrightarrow RC_2R'Co_2(CO)_6 + 2\,CO$$

At low temperatures the equilibrium is nearly quantitatively shifted to the right side and the inhibition in a start-up operation or discontinuous hydroformylation, even with a single addition of acetylene, may last hours until at last all the acetylene has been hydrogenated [145]. At high carbon monoxide pressure and especially at higher temperatures the inhibition effect is smaller (see fig. 4).

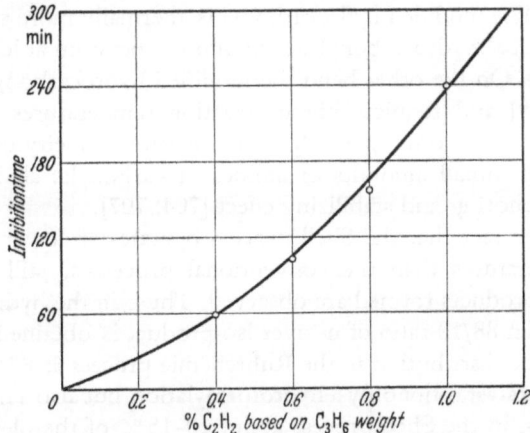

Fig. 4 [45]. Dependence of the inhibition time in the hydroformylation of propylene at 150 °C ± 2 ° on the amount of acetylene added

In continuous propylene hydroformylations, up to 0.1 wt-% of propyne and up to 0.5 wt-% of allene in the propylene feed may be tolerated normally.

A number of metals and their ions such as lead, mercury, bismuth and zinc [27, 147] were also reported to hinder not only the hydroformylation but also the homogeneous hydrogenation of the aldehydes formed to the alcohols.

It may be assumed that this hindrance is due to formation of carbonyls containing two or more different metal atoms, which are less active than the pure hydrocarbonyls of Co, Rh and Ru [147].

3.6. Catalyst Modifiers

For many years it has been a well established fact that one or more carbonyl ligands in the carbonyls or hydrocarbonyls of Co, Rh and Ru may be replaced by other ligands. Such modified carbonyls were early suggested to be suitable catalysts for carbonylation or hydroformylation reactions [922–926].

Suitable ligands include nitriles [990], tertiary amines, phosphines, phosphites, other compounds with trivalent phosphorus, arsines and stibines.

Especially the carbonyls modified by one or more molecules of the listed phosphorus compounds [942, 945, 946, 1037, 1044, 1055] are of interest since they exhibit a number of interesting properties which have led to their industrial application (Shell oxo process, see also chapter on technical hydroformylations, page 71) [148, 701–707, 743, 790, 791, 966, 977, 978]. In the Shell process trialkylphosphines are fed to the oxo reactor along with Co-salts. The phosphine cobaltcarbonylhydride ($R_3PCo(CO)_3H$) is formed in situ. Compared to the application of unmodified hydrocarbonyls, the following differences arise. The modified hydrocarbonyl is thermally more stable than the unmodified cobalthydrocarbonyl which allows operation at lower pressures (35–100 atm). On the other hand the modified hydrocarbonyl is less reactive [932, 939] and requires higher reaction temperatures (180–200 °C) and even then gives only $1/_5$ to $1/_6$ of the reaction velocity of the conventional method. Small amounts of amines or carboxylic acids are claimed to have a promoting and stabilizing effect [704, 707].

Despite the fact that the Shell process operates at lower pressure and higher temperatures than the conventional processes, still higher n/iso ratios of the products formed are observed. Thus, in the hydroformylation of propylene an 88/12 ratio of n- over iso-product is obtained, whereas for comparison the distribution in the Ruhrchemie process is 80/20. This type of modified catalyst is not only a hydroformylation but also a hydrogenation catalyst. Thus, in the Shell process about 10–15% of the olefin fed is lost through hydrogenation to the paraffin whereas the figures for the conventional oxo processes are only about 2–3%.

The aldehydes formed in the Shell process are nearly quantitatively hydrogenated in the oxo reactor to the alcohols. For a detailed comparison of both processes see reference 927.

If triphenylphosphite [149] or some other phosphorus compounds [928, 1037] with a more complicated structure [928] are used as catalyst modifiers, no such hydrogenation is observed but high yields of aldehydes are obtained.

Interesting results with modified catalysts were published recently by Wilkinson *et al.* [929, 930, 933] who succeeded in reacting olefins to aldehydes at such mild reaction conditions as 25 °C and 1 atm using $HRh(CO)(PPh_3)_3$. The ratio of n- over iso-aldehydes formed is about 20 and thus much higher than in the conventional or in the Shell process.

Nearly at the same time R. L. Pruett and J. A. Smith [931] showed that high ratios of n- over iso-aldehydes can be obtained in olefin hydroformylation with very similar catalysts of the type $HRh(CO)(P(OR)_3)_3$ where R may be alkyl or aryl.

The differences in the behavior of modified and unmodified hydrocarbonyls have been extensively discussed during the last two years and mechanisms have been proposed [892, 911, 929, 931–944, 984, 1037].

The increased thermal stability of the modified hydrocarbonyls can easily be explained. The trivalent phosphorus ligands are better σ-donors than CO but poorer π-acceptors [911, 934, 947]. As a consequence the remaining CO-ligands are more strongly bonded [942] since the transition metal atom has the tendency to transfer the increased negative charge obtained from the phosphorus to the CO-ligands through π-back donation [911, 934]. Some of this increased negative charge will also be transferred to the hydrogen atom giving this a more hydridic character than in the cobalt hydrocarbonyl [911, 932, 934, 948].

It is also well established that the resulting modified hydrocarbonyl is a trigonal bipyramid (dsp^3-hybrid), the three CO-ligands being in the equatorial position and hydrogen and phosphorus in the axial position [911]. The whole molecule is more bulky than cobalt hydrocarbonyl.

The increased strength in the bonding of the CO-ligands is the reason for the higher thermal stability and the lower reactivity of the modified hydrocarbonyls, since dissociation of the hydrocarbonyl will be lower than in cobalt hydrocarbonyl.

Three different arguments have been put forward to explain the increased stereoselectivity of the phosphine-modified catalysts:

a) The increased hydridic nature of the hydrogen atom in the modified hydrocarbonyl which should lead to a preferred addition of this hydrogen to the more electropositive nonterminal C-atom, whereas the $Co(CO)_3PR_3$ moiety should preferably be added to the terminal C-atom [911, 934, 944].

b) Decrease in olefin isomerization prior to hydroformylation which should be due to a lower isomerization activity of the modified hydrocarbonyl [934].

c) Increased steric hindrance in the ligand field of the modified hydrocarbonyl through the more bulky phosphorus ligand resulting in hindered formation of branched alkyl-$Co(CO)_3PR_3$ [911, 912, 944].

It has been demonstrated that olefin isomerization is indeed slower when using the modified hydrocarbonyl [912]. However, this hardly is of any influence on the isomer distribution. As can be seen from the results of Hershman and Craddock [936], of Kniese, Nienburg and Fischer [912] and of Fell, Rupilius and Asinger [943], nearly the same aldehyde isomer distribution is obtained when reacting terminal and internal olefin isomers. This is in line with the results of Pino, Pucci and Piacenti [89] who showed that olefin isomerization prior to hydroformylation hardly ever occurs under normal oxo conditions. The isomerization very likely does proceed through intermediate alkyl- or acylcobalt trialkylphosphine carbonyls. Thus, it looks as if only a) and c) are the determining factors in the high stereoselectivity.

"a)" is reasonably obvious and generally accepted.

However, "c)" needs some more explanation since it might not be clear on the first look. If the mechanism with the modified hydrocarbonyl is the same as the one proposed by Heck and Breslow for the conventional cobalt hydrocarbonyl, then the trigonal bipyramid (A) (dsp^3 hybrid) should dissociate in a first step to give a tetrahedron (B) (sp^3 hybrid)

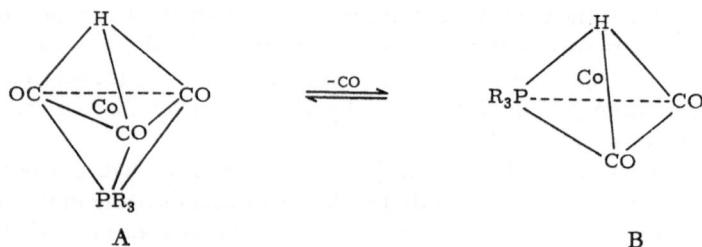

Since (B) is more bulky than $HCo(CO)_3$ both the π-complex formation with the olefin and the formation of the alkylcobalt compound will be more hindered. Thus, the least hindered product, namely the terminal alkylcobalt compound, will be formed. However, taking into account the high stability of the $HCo(CO)_3PR_3$, it may well be that a different mechanism is involved and that no dissociation to $HCo(CO)_2PR_3$ occurs, but a direct addition of the hydrocarbonyl across the olefinic double bond occurs to yield $R-Co(CO)_3PR_3$ in one step [911, 934, 935, 944].

There would be no increased steric hindrance in this addition since the phosphine is in the axial position in (A) and the alkyl group will take the position which was held by hydrogen before. But the steric hindrance will arise when the bipyramid (C) rearranges to the tetrahedron

(D), that is in the formation of the acyl compound. Here again the straight chain compound will be preferentially formed due to the bulky phosphine ligand.

C   D

Effects different from those observed with modified cobalt hydrocarbonyl were found with modified rhodium hydrocarbonyls. Thus, Fell, Rupilius and Asinger [943, 1038, 1053] found that isomerizations are nearly completely suppressed when working with rhodium carbonyls and excess trialkylphosphines and almost only aldehydes were found, originating from an addition of the formyl group to the C-atom which had previously formed the carbon double bond.

The mechanisms of the hydroformylations carried out by Wilkinson *et al.* [929, 933] and Pruett and Smith [931] with rhodium carbonyls in the presence of tertiary phosphines are still under discussion [931, 949, 950, 1039]. It is not yet clear whether the well-defined HRhCO(PR$_3$)$_3$ or rather the HRh(CO)$_2$(PR$_3$)$_2$ [929] is the actual catalyst.

Moreover, it cannot be decided from the data available so far whether a dissociative mechanism (similar to that proposed by Heck and Breslow for the hydroformylation with HCo(CO)$_4$ according to scheme 1) or rather an associative attack of the alkene on the hydrocarbonyl according to scheme 2 [929, 1039] is followed (see page 26).

In view of the mild reaction conditions, scheme 2 looks quite attractive. A similar scheme has also been suggested for the hydroformylation with phosphorus-modified cobalt hydrocarbonyls [911].

Anyhow, it seems reasonably clear that strong steric hindrance in the above complexes is mainly responsible for the preferential formation of branched aldehydes. Besides that, electronic factors may also play a certain role.

3.7. Separation and Recovery of Oxo Catalysts

The separation and recovery of the oxo catalysts are very essential steps in the process. The ease and completeness of this operation is of great economic importance to oxo production units. Cobalt hydrocarbonyl and dicobalt octacarbonyl are not only highly soluble in organic solvents but are also very volatile and are taken out of a solution with gas streams. This

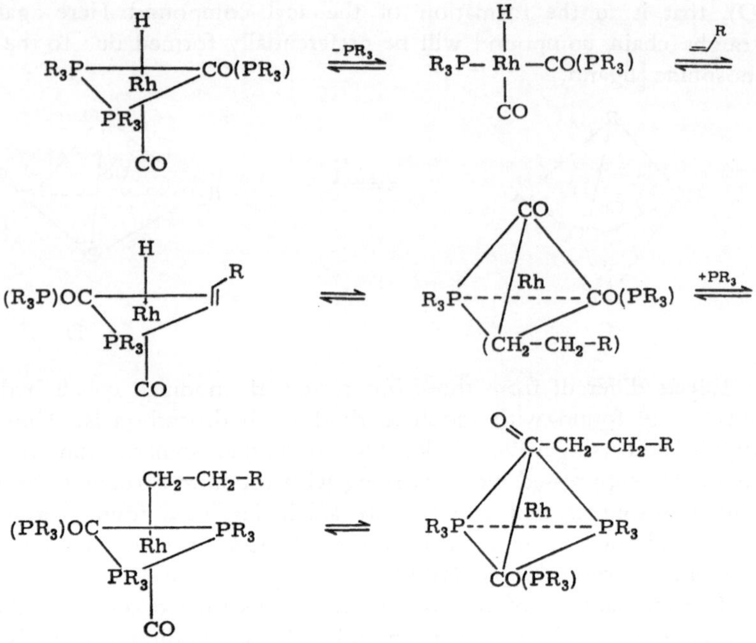

$$R_3P\text{—}\underset{\underset{CO}{|}}{\overset{\overset{H}{|}}{\underset{Rh}{|}}}\text{—}CO(PR_3) \quad \underset{\Longleftarrow}{\overset{-PR_3}{\longrightarrow}} \quad R_3P-\underset{\underset{CO}{|}}{\overset{\overset{H}{|}}{Rh}}-CO(PR_3) \quad \rightleftharpoons$$

$$(R_3P)OC\text{—}\underset{\underset{CO}{|}}{\overset{\overset{H}{|}}{\underset{Rh}{|}}}\text{—}{\parallel}^{R} \quad \rightleftharpoons \quad R_3P\text{---}\overset{CO}{\underset{(CH_2-CH_2-R)}{Rh}}\text{---}CO(PR_3) \quad \overset{+PR_3}{\rightleftharpoons}$$

$$(PR_3)OC\text{—}\underset{\underset{CO}{|}}{\overset{\overset{CH_2-CH_2-R}{|}}{\underset{Rh}{|}}}\text{—}PR_3 \quad \rightleftharpoons \quad R_3P\text{---}\overset{O=C-CH_2-CH_2-R}{\underset{CO(PR_3)}{Rh}}\text{---}PR_3$$

Scheme 1

$$R_3P\text{—}\underset{\underset{CO}{|}}{\overset{\overset{H}{|}}{\underset{Rh}{|}}}\text{—}CO(PR_3) \quad \overset{R}{\underset{\Longleftarrow}{\rightleftharpoons}} \quad \begin{array}{c} R_3P\text{—}\overset{\overset{H}{|}}{\underset{Rh}{|}}\text{—}{\parallel}^{R} \\ R_3P\text{—}\underset{\underset{CO}{|}}{|}\text{—}CO(PR_3) \end{array}$$

$$\longrightarrow \quad R_3P\text{—}\underset{\underset{CO}{|}}{\overset{\overset{CH_2-CH_2-R}{|}}{\underset{Rh}{|}}}\text{—}CO(PR_3)$$

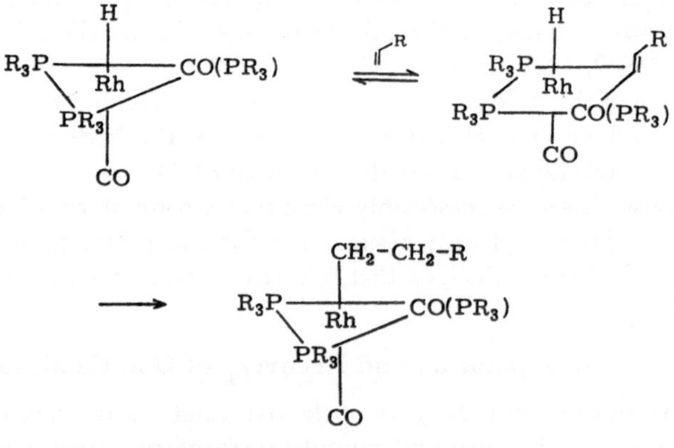

Following steps as in scheme 1

Scheme 2

volatility may lead to metal deposits in parts of the plant. On the other hand cobalt catalyst residues remaining in the demetallized raw oxo product cause condensation and oxidation reactions in the distillation of the reaction products and result in colored products in the work-up. Therefore a quantitative separation and recovery of the catalyst is necessary.

There are a number of different methods applied. Some of these cannot be recommended for plant operation but only for laboratory work.

a) Separation of oxo product from catalyst by distillation.

In case of low boiling aldehydes (propionaldehyde, butyraldehyde, etc) and carbonyls of low volatility (Rh-carbonyls, phosphine modified cobalt or rhodium catalysts), the aldehydes may be separated from the reaction product by flash distillation [150]. In this operation the metal carbonyls remain in the bottoms. Depending on temperature, length of treatment and stability of the carbonyls, varying amounts of them are decomposed. The corresponding metals formed also remain in the bottoms. This method is limited to laboratory operations (with the exception of phosphine modified catalyst — see chapter on modified catalysts). It cannot be recommended for cobalt carbonyls since they are too volatile.

b) Thermal decomposition of the carbonyls and subsequent separation of organic material and catalyst decomposition products.

This method was already applied in the first technical oxo reaction unit in the so-called "two-tower-process" (hydroformylation reactor and decobalting reactor) in which both reactors were filled with cobalt on carrier; the first operating with high, the second with low pressure. In the first reactor the solid cobalt was converted into hydrocarbonyl, in the second reactor the dissolved carbonyls were decomposed (because the pressure was too low to keep them stable) and precipitated on the carrier. When most of the cobalt had been transferred from the first to the second reactor the functions of the two reactors were reversed [151, 152].

The method may also be applied on the laboratory scale. In this case pressure is released at the end of the reaction from the reaction vessel — e. g. an autoclave — and the gas vented while heating is continued, resulting in decomposition of the carbonyls and precipitation of the corresponding metals. The big disadvantage of this type of catalyst removal is the fact that the metal not only decomposes on the carrier but also on the wall of the reactor, thus reducing heat transfer through the wall and clogging the reactor.

A way out of these difficulties was found by introducing the heat necessary for catalyst decomposition by means of a liquid, e. g. a recycle of demetallized hot reaction product, distillation residues, hot water or steam [101, 735, 755, 763; 764]; oxygen or air may also be applied simultaneously [960].

When these means are applied, cobalt is precipitated in powder form — provided the walls of the reactor are kept at a temperature low enough to avoid decomposition on it. Inert gases are often introduced along with the hot media for faster removal of the CO resulting from decomposition of the carbonyls and acceleration of the decomposition. Solid catalyst decomposition products may be separated from the resulting mixture by conventional methods such as filtration and the organic material purified by distillation.

c) Decomposition of the carbonyls by hydrogenation followed by separation of organic material and catalyst decomposition products [101, 153].

In this method hydrogen is introduced into the hot reaction mixture and the carbonyls are decomposed into metal and carbon monoxide by hydrogenation.

Carriers on which the metal is precipitated may be used as in method b). This method can be recommended if alcohols rather than aldehydes are the desired reaction products.

d) Separation of catalyst by treatment of the reaction product with chemicals. The oxo catalyst may be extracted by acids from the crude oxo product after pressure reduction. Oxidation of the carbonyls by introducing oxygen or air may be applied simultaneously [961, 962].

The following acids are recommended for this type of catalyst removal: sulfuric acid [9, 765], carbon dioxide [154, 155] and carboxylic acids such as oxalic acid [156, 766, 767], formic acid [157–160, 768, 769, 780] acetic acid [160, 769–772], propionic acid, fumaric acid and maleic acid [775, 776].

Cobalt hydrocarbonyl may also be extracted with alkali [161, 162]. In this procedure the hydrocarbonyl may be regenerated from the alkaline solution by acidification with mineral acid and recycled to the oxo reactor in gaseous form.

Another method was reported by G. Noyori *et al.* who introduced chlorine into the crude reaction product and removed the cobalt chloride formed by filtration or extraction with water [762].

Special care has to be taken if precious catalysts such as rhodium carbonyls have to be recovered, and special methods have been worked out for this purpose [958, 959].

Besides the classical methods of metal analysis, IR spectroscopy may be applied advantageously for determination of metal-carbonyl residues in organic material, since carbonyls show a strong band in the carbonyl-region, which may be used for continuous control of catalyst removal from the reaction product.

e) Recovery of volatile carbonyls from off gases.

The carbonyls applied as oxo catalyst are (with some exceptions) highly volatile and may thus be carried out of the reaction systems with the off gas. It may therefore in many cases be useful to wash the off gas either with fresh olefin or with solvents or oil.

4. Influence of Pressure and Temperature

As mentioned above, hydroformylation reactions occur under atmospheric pressure at normal temperature with stoichiometric amounts of cobalt carbonyls. However, with catalytic amounts of cobalt catalysts a minimum CO partial pressure is necessary for reformation and stability of $Co_2(CO)_8$, or $HCo(CO)_4$, as the case may be (see page 15). A small increase of the CO partial pressure above this value first results in an increase of the reaction velocity until a maximum is reached depending on temperature and olefin structure. However, further increase of the CO-partial pressure causes a decrease in the reaction velocity [38, 40, 120], (see also section on reaction mechanism).

Increase of the hydrogen partial pressure always leads to increased reaction velocity [38, 40, 120]. At very high pressures this effect becomes smaller [194].

In the pressure range usually applied, the empirical equation which was formulated by G. Natta *et al.* based on kinetic measurements may be used in a good approximation.

$$\frac{d \text{ (aldehyde)}}{dt} = k \text{ (olefin) (Co) } (p_{H_2}) (p_{CO})^{-1}$$

With equimolar amounts of CO and H_2, the hydroformylation reaction seems to be independent of pressure over a wide range, due to the opposing effects mentioned above.

The observed dependence on the total pressure in continuous processes can be related to an insufficient diffusion of gas- and liquid-phase [195, 774].

Generally, pressures of 80 to 300 atm are applied at reaction temperatures above 100 °C, with the exception of the Shell process which has been successfully applied in the last few years [148]. In the Shell process cobalt carbonyls modified by phosphine ligands are used as catalysts. They exhibit a high thermal stability and allow operations at pressures of 3 to 35 atm even at such high temperatures as 180—200 °C.

Similar to a minimum CO partial pressure, a minimum temperature is also required if catalytic amounts of Co catalysts are to be applied. The minimum temperature depends on olefin structure and diffusion of the reactants [756]. In the case of olefins, minimum reaction temperatures generally lie between 50 and 80 °C.

Usually a rapid reaction can be achieved at temperatures between 90 and 200 °C. Generally, increase of the temperature causes an increase of the reaction velocity if care is taken to maintain a pressure which guarantees the stability of the catalyst. In case of butene-1, e. g., the reaction velocity increases by a factor of 100 for a temperature increase from 90 to 140 °C [126].

Table 4 [126]. *Temperature influence on the reaction velocity*

Butene-1, 0.05 wt-% of catalyst based on butene-1, 246 atm CO/H_2 (1:1)

Reaction temperature (°C)	Relative reaction velocity
90	0.01
100	0.04
120	0.20
140	1.00

On the other hand, higher reaction temperatures often favor side reactions, especially the hydrogenation of the starting material towards the saturated hydrocarbons, and of the aldehydes formed in the hydroformylation to the corresponding alcohols.

Pressure and temperature exhibit a strong influence on the isomer distribution of the hydroformylation products. Thus, in the case of straight chain olefins, higher CO partial pressure favors the formation of straight chain aldehydes at constant reaction temperatures [62, 196].

Table 5 [62]. *Hydroformylation of propylene in presence of ethyl orthoformiate*

$HC(OC_2H_5)_3 = 90$ g, $C_2H_5OH = 80.5$ g, $Co_2(CO)_8 = 1$ g, $p_{H_2} = 80$ atm

Temp. (°C)	p_{CO}	% Straight chain isomer	Temp. (°C)	p_{CO}	% Straight chain isomer
108	4	45.4	90	12	60.3
108	6.5	49.7	90	25	64.8
108	9	51.4	90	70	69.5
108	10	53.4	90	113	70.0
108	12.5	60.6	90	174	70.2
108	21	63.5			
108	31	65.0	80	11	63.1
108	66	70.2	80	15	65.2
108	104	72.7	80	30	67.5
108	147	73.3	80	48	67.7
108	224	73.4	80	70	67.7

In the lower pressure range this influence is remarkably strong. However, it becomes smaller at higher CO partial pressures.

According to S. Brewis, the amount of straight chain products drops if very high pressures are applied (see fig. 5).

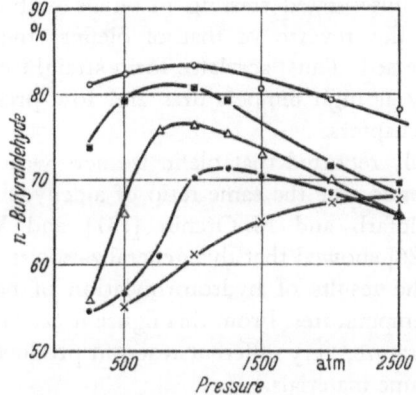

Fig. 5 [198]. n-Butyraldehyde in the reaction products of propylene hydroformylation in dependence on overall pressure

o = 100 °C, ■ = 130 °C, ▽ = 160 °C, ● = 200 °C, x = 250 °C

The influence of the temperature can be seen from fig. 6.

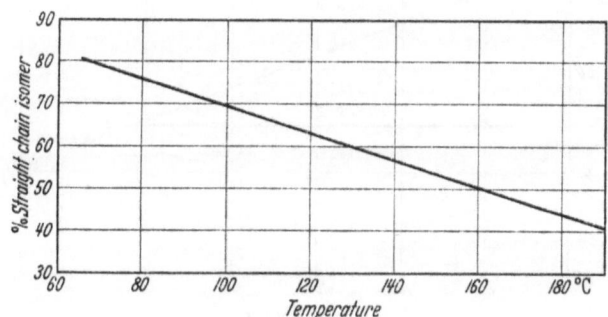

Fig. 6 [126]. Hydroformylation of heptene-1 at various temperatures
Olefin: heptene-1 with 8% of heptene-2: 0.5 wt-% Co/olefin;
$H_2/CO = 3/1$; overall pressure 246 atm

According to F. Piacenti higher hydrogen partial pressure also favors
the formation of straight chain products although the effect is much smaller
than that of carbon monoxide [1051].

At constant pressure a decrease of reaction temperature results in a
remarkably higher yield of straight chain aldehydes. Obviously, high CO
pressure, extensive mixing both resulting in high CO concentration in the
liquid phase and low temperature favor the formation of terminal alkyl
cobalt carbonyls and result in high yields of straight chain compounds.
Contrariwise, at low CO pressure or high temperature more alkyl tricarbonyl and accordingly more branched aldehydes are formed.

In the case of unsaturated starting materials with an electron configuration which is the reverse of that of olefins, the above-mentioned effects are also reversed. Thus, acrylates form straight chain reaction products preferentially at high temperatures and low pressures. For details see the individual chapters.

It was repeatedly reported that olefin isomers with non-terminal and terminal double bonds give the same ratio of aldehyde isomers on hydroformylation I. Goldfarb and M. Orchin [131] and V. L. Hughes and I. Kirshenbaum [126] showed that this generalization is incorrect.

Fig. 7 shows the results of hydroformylation of heptene-1 and heptene-2 at various temperatures. From this figure it can be seen that, depending on the temperature, very different reaction products can be obtained starting with the same materials.

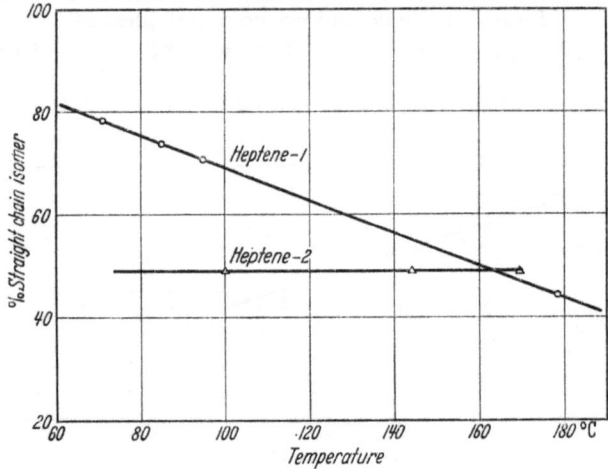

Fig. 7 [126]. Reaction product distribution in the hydroformylation of terminal and non-terminal olefins at various reaction temperatures
$H_2/CO = 3/1$, overall pressure 246 atm

Obviously, the isomerization according to the scheme on page 11 is slower than the hydroformylation at low temperatures. Thus, at these low temperatures higher amounts of straight chain aldehydes are obtained from α-olefins than from their internal isomers.

The findings of other authors that straight chain olefins, with both terminal and non-terminal double bonds, always form the same ratio of isomeric aldehydes can be explained by the fact that they limited their experiments to reaction temperatures of 140 to 170 °C. In this range, indeed, nearly the same isomer distributions are obtained from both types of olefins, as can be seen from fig. 7.

5. Solvents of the Hydroformylation Reaction

Suitable solvents, or diluents, in the hydroformylation reaction are: aliphatic, cycloaliphatic and aromatic hydrocarbons, aliphatic, cyclic and aromatic ethers, aliphatic alcohols, nitriles, anhydrides, ketones, esters, lactones, lactams, orthoesters and water.

In many cases the reaction products themselves, or recycled high boiling residues, are used as diluents. In technical operations the reaction is generally run without diluents and solvents are used only for charging the catalyst.

The solvent influence on the reaction velocity of the hydroformylation was studied in detail [199–201]. The influence varies, depending on the unsaturated starting material.

While I. Wender et al. [199] in the case of cyclohexene observed differences in the reaction velocity only up to a factor of 1.5 with different solvents, Iwanaga et al. [200] and also Wakamatsu et al. [201] stated a 6 times higher reaction velocity in the conversion of acrylonitrile with methanol as solvent compared to benzene.

In case of acrylates even differences of a factor of 7 were observed with various solvents (see table 6).

Generally primary and secondary alcohols allow higher reaction velocities than other solvents. The reason may be a faster formation of hydrocarbonyls from the metal carbonyls in the presence of alcohols. Alcohols, however, are not generally suited as solvents in the hydroformylation reaction, especially not at low reaction temperatures. In this case not the aldehydes but the corresponding acetals are obtained as reaction products [174, 202–204, 254].

However, in some cases the formation of acetals may be desired if sensitive and valuable aldehydes are to be trapped in the form of their acetals. Acetals are formed in high yields using orthoesters as solvents [205]. With acetic acid anhydride as solvent aldehyde-acetates are the reaction products.

According to Hagemeyer et al. [206] higher n/iso ratios of aldehydes are achieved in the hydroformylation of olefins using orthoesters as solvents.

Y. Takegami et al. [858] studied the effects of other solvents in this respect. They took toluene as a standard solvent in the propylene hydroformylation (n-aldehyde 77–73%). Compared to toluene, dioxane and esters such as butylacetate e.g. gave higher n/iso-ratios (n-aldehyde 83 to 79%) while acetone and diethyl ether increased the formation of the branched isomer (n-aldehyde 66–60%).

As mentioned above, water can also be used as diluent [109, 199, 208–211, 214], if needed, in the presence of surface active compounds

Table 6 [200]. *Influence of solvent on the reaction velocity in hydroformylate of methacrylate*

120 °C, 260 atm, $CO/H_2 = 1:1$

Solvent	Reaction velocity constant $(k_a \cdot 10^3 \text{ min}^{-1})$
Benzene	41.8
Toluene	43.4
Ethyl acetate	36.4
Ethyl ether	41.2
Methanol	157.0
Ethanol	186.0
n-Butanol	167.0
Cyclohexanol	154.0
tert. Butanol	66.2
Acetone	59.5
Ethyl methyl ketone	39.1
Tetrahydrofuran	57.1
Dioxane	27.4
Acetic acid anhydride	30.2
Acetonitrile	77.2

[212–213]. The addition of water to organic solvents in the hydroformylation on the one hand suppresses undesired side reactions, such as aldehyde condensation reactions; however, on the other hand it decreases the reaction velocity (see fig. 8).

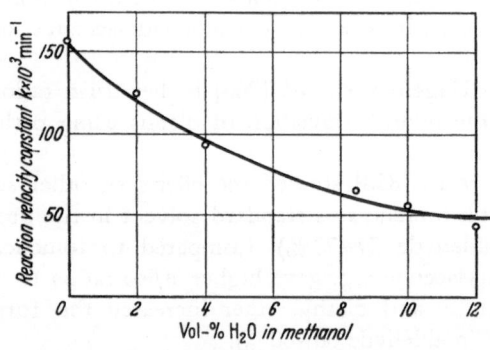

Fig. 8 [215]. Influence of water content on reaction velocity of hydroformylation of methyl acrylate in the presence of methanol as solvent

6. Hydroformylation of Various Structures

6.1. Olefins

In principle, all olefins participate in the hydroformylation reaction; however, their reactivities vary markedly. I. Wender *et al.* [199] made a systematic investigation of the reaction rate as a function of the olefin structure and found a variation of a factor of 50 (see table 7).

Table 7 [199]. *Hydroformylation of olefins at 110 °C*

Reaction conditions: 0.5 mole olefin; 65 ml methylcyclohexane solvent;
2.8 g (8.2 × 10^{-3} mole) dicobaltoctacarbonyl;
CO/H_2 = 1:1; initial pressure at room temperature: 233 atm

Class of olefin	Specific reaction rate constant * ($10^3 k$ min^{-1})
(A) Straight chain, terminal	
Pentene-1	68.3
Hexene-1	66.2
Heptene-1	66.8
Octene-1	65.6
Decene-1	64.4
Tetradecene-1	63.0
(B) Straight chain, internal	
Pentene-2	21.3
Hexene-2	18.1
Heptene-2	19.3
Heptene-3	20.0
Octene-2	18.8
(C) Branched, terminal	
4-Methylpentene-1	64.3
2-Methylpentene-1	7.32
2,4,4-Trimethylpentene-1	4.79
2,3,3-Trimethylbutene-1	4.26
Camphene	2.2
(D) Branched, internal	
4-Methylpentene-2	16.2
2-Methylpentene-2	4.87
2,4,4-Trimethylpentene-2	2.29
2,3-Dimethylbutene-2	1.35
2,6-Dimethylheptene-3	6.23
(E) Cyclic	
Cyclopentene	22.4
Cyclohexene	5.82
Cycloheptene	25.7
Cyclooctene	10.8
4-Methylcyclohexene-1	4.7

* Except for camphene and cyclooctene, the values were determined in duplicate. The error, determined by a statistical analysis of 55 experiments, was ± 1.5%

The most reactive, with cobalt catalysts, were straight chain, terminal olefins, depending only slightly on molecular weight. Even terminal double bonds in high polymers react similarly [781, 964, 976].

With straight chain internal olefins, the rate is approximately one third that of the corresponding terminal olefins. The position of the internal double bond is not important, e. g., heptene-2 and heptene-3- react at practically the same rate.

Branching of the olefin structure always decreases the reaction rate. The strongest decrease occurs when one of the carbon atoms of the double bond is substituted. Branching at more distant carbons gives a less marked, but still significant reduction in reactivity. The slowest rates occur with internal, double bond substituted olefins.

Wakamatsu [172] has reported that, with rhodium catalysts, internal olefins react faster than terminal ones. Cyclic olefins gave conflicting results. Cyclohexene reacted very slowly, while cyclopentene and cycloheptene were even faster than internal, straight chain olefins. The increased rates result from the strain in these olefins, the relief of this ring strain increasing the reactivity of the double bond. Contrary to Wakamatsu's findings, Heil and Marko [956] report the following order of reactivity with rhodium carbonyls and various structures of olefins: styrene $\gg$ linear α-olefins $>$ linear internal olefins $>$ single branched olefins $>$ multiple branched olefins.

Symmetrical and fluxional olefins give consistent products (table 8).

Table 8. *Hydroformylation of symmetrical and fluxional[a] olefins*

$CO/H_2 = 1:1$

Olefin	Pressure (atm)	Temp. (°C)	Product	Yield (%)	Ref.
Ethylene	40—300	135	Propionaldehyde	99	[216, 987]
Cyclopentene	200—300	120—125	Cyclopentanealdehyde, Cyclopentylcarbinol	65	[217]
Cyclohexene	250	120	Cyclohexanealdehyde, Cyclohexylcarbinol	~100	[105]
Cyclooctene	200	120—180	Cyclooctylcarbinol		[217]
Camphene	210	125—150	Homoisocamphenilan-aldehyde	65	[217]

[a] Structures which are unchanged after double bond isomerization

In all other cases, one obtains a mixture of aldehyde isomers. The composition of the isomeric products depends greatly on the structure of the starting olefin and the reaction conditions (see chapters 3 and 4). Table 9 gives several examples of products obtained from olefins of various structures.

Table 9. *Hydroformylation of unsymmetrical or isomerizable olefins*

Olefin	Products	Yield (%)	Reference
Propylene	Butyraldehyde + isobutyraldehyde	80	[21, 105, 168, 218–228]
Butene-1	N- and isovaleraldehyde	96.8	[105]
Butene-2	Isovaleraldehyde	96	[113]
Isobutylene	3-Methylbutane-1-ol + neopentylalcohol	82	[25, 229, 230, 237]
Pentene-1	Hexanol + 2-methylpentanol + 2-ethylbutanol (5 : 4 : 1)	—	[25]
	C_6-Aldehydes and alcohols	92	[743]
Pentene-2	C_6-Aldehydes	75	[24, 29]
2-Methylbutene-1	4-Methylpentanol + 3-methylpentanol + 2,3-dimethylbutanol (11 : 9 :1)	—	[25]
2-Methylbutene-2	4-Methylpentanol + 3-methylpentanol + 2,3-dimethylbutanol	—	[25]
3-Methylbutene-1	4-Methylpentanol + 3-methylpentanol + 2,3-dimethylbutanol	—	[25]
2,3-Dimethyl-butene-1 + 2,3-dimethyl-butene-2	3,4-Dimethylpentanol	—	[25]
3,3-Dimethyl-butene-1	4,4-Dimethylpentanol	—	[25]
2-Ethylbutene-1	3-Ethylvaleraldehyde	55	[24]
Hexene-1	n-Heptaldehyde + 2-methylhexaldehyde	90	[178]
2-Methylpentene-1	C_7-Aldehydes	77	
2-Methylpentene-3	5-Methylhexanol + 3-methylhexanol + 2,4-dimethylpentanol (4 : 3 : 3)	—	[25]
(+) (S)-3-Methyl-pentene-1	4-Methylhexanal	92,1	[89]
Isoheptene	Isoöctanols	74.6	[9]
Octene-1	Nonanol (75.6 %) + isononanol	85.3	[232]
cis-Octene-4	Nonanol (53.9 %) + isononanol	78.4	[232]
trans-Octene-4	Nonanol (57 %) + isononanol	85.5	[232]
Dibutylene	3-Ethylheptanol	39	[233]
2-Ethylhexene-1	C_9-Aldehydes	23	[223]
Diisobutylene	3,5,5-Trimethylhexanol	90	[25, 234, 235, 733]

Table 9 (continued)

Olefin	Products	Yield (%)	Reference
Styrene	Methylphenylacetaldehyde Methylphenyl acetaldehyde + hydrocinnamaldehyde	30 91 *	[24, 29, 205, 226, 1052]
α-Methylstyrene	Aldehyde + isopropylbenzene	—	[173]
α-Pinene	2- or 3-Formyl-2,6,6-trimethyl-bicyclo-(3.1.1)-heptane	—	[205]
1-Vinylnaphthalene	Methyl-(1-naphthyl)-acetaldehyde	29	[24]
Octadecene	C_{19}-Aldehydes	54	[24, 756]

* Rhodium catalyst

Hydroformylation of linear terminal olefins under the standard conditions of the technical Oxo process (90–150 °C, 100–300 atm, cobalt catalyst) give linear aldehydes (50—80%), together with a smaller amount of branched aldehydes.

Isomers of linear internal olefins give practically the same product distribution at higher temperatures. However, at lower temperatures terminal olefins give a higher proportion of linear aldehydes than internal olefins (see fig. 9, p. 49).

Irregularities occur with branched olefins. As Nienburg et al. [236] and Keulemans et al. [25] first pointed out, no quaternary carbon atoms are formed by olefin hydroformylation, at least under the standard conditions given above; that is, the formyl group is not attached to a carbon atom that is branched. Accordingly, isobutylene forms almost exclusively 3-methylbutanal [237], only about 5% of the isolated product being pivalaldehyde.

$$H_3C \diagdown C=CH_2 + CO/H_2 \longrightarrow H_3C \diagdown CH-CH_2-CHO \quad 95\% \atop H_3C \diagup \qquad \qquad \qquad H_3C \diagup$$

$$H_3C-\underset{\underset{CH_3}{|}}{\overset{\overset{CH_3}{|}}{C}}-CHO \qquad \qquad 5\%$$

Tetramethylethylene gives almost exclusively 3,4-dimethylpentanal. Furthermore, branching in the vicinity of a vinylic carbon atom hinders the building of a formyl group on this carbon [25].

In general, the formation of many aldehyde isomers is undesirable for commercial production. Much research has therefore been directed to variations of the reaction conditions to improve the yield of individual products. In particular, higher yields of linear aldehydes — precursors to the industrially important plasticizer and detergent alcohols — were desired. These investigations have been successful. It is possible to vary markedly the proportion of individual isomers through variation of pressure and temperature.

Table 10 [126]. *Influence of pressure on isomer distribution*
Heptene-1, 0.5 wt-% Co/olefin, 85 °C, CO : H$_2$ = 1 : 3

Total pressure, atm	42	247	352
% Linear products	45	75	75
% Branched products	55	25	25

Higher carbon monoxide partial pressure (see table 5, p. 30), in some cases higher total pressure (table 10) and lower temperature (table 11) all lead to higher yields of linear products.

Table 11 [126]. *Influence of temperature on isomer distribution*
Pressure: 247 atm

Olefin	Temp. (°C)	Ratio CO:H$_2$	% Linear products (I)	% Branched products (II)	Ratio I:II
Butene-1	70	1 : 3	73	27	2.7 : 1
	90	1 : 3	71	29	2.4 : 1
	140 c	1 : 1	60	40	1.5 : 1
	180 c	1 : 1	45	55	0.8 : 1
Heptene-1 a	70	1 : 3	80	20	4.0 : 1
	85	1 : 3	75	25	3.0 : 1
	100	1 : 3	72	28	2.6 : 1
	180 b	1 : 3	43	57	0.75 : 1

a Contained 8% heptene-2; 0.5 wt-% Co/olefin
b 0.05% w Co/Olefin
c 0.08% w Co/Olefin

Especially important has been the use of certain phosphines, as for instance, in the Shell process (see p. 22). Over 85% linear products are obtained at total pressures as low as 30 atm. However, overall alcohol yield is lowered. A certain amount of paraffin is also formed through hydrogenation of the starting olefin.

If branched products are desired, it is preferable to use appropriate internal olefins (see fig. 7, p. 32) and work with lower CO partial pressures and higher temperature. It is also possible to use rhodium catalysts instead of cobalt, or mixed Rh/Co catalysts, which generally result in higher yields of branched aldehydes [105, 106, 172–179] (table 12).

Table 12 [106]. *Effect of rhodium catalyst on isomer distribution*

Butene-1, 245 atm CO/H_2 (1:1), rhodium as sesquioxide, cobalt as preformed carbonyl, 105 °C

| Mole % Catalyst | | % Branched | Relative |
Rh	Co	products	rate
0.05	—	85	1.0
—	1.6	30	0.15
0.005	1.6	80	0.85

Efforts to find conditions where the formyl group will be attached to a branched carbon atom have met with only limited success. With iso-butylene at higher temperature (220 °C) and pressure (425 atm) required for catalyst stability, an 8% yield of neopentyl compounds is obtained, together with 61.6% yield of isopentyl compounds. At the same time, 25% of the starting material was hydrogenated [237]. Good yields of the neo-pentyl structures have been obtained only at room temperature with stoichiometric quantities of hydrocarbonyl [35] (see also p. 9).

A reaction closely related to olefin hydroformylation is the Reppe alcohol synthesis [239, 1003]. It is carried out in alkaline media with $Fe(CO)_5$ catalyst and differs from the Roelen reaction in that water is used instead of hydrogen, with the result that alcohols are obtained directly at lower temperatures. The active catalyst is believed to be a salt of the iron hydrocarbonyl.

$$R{-}CH{=}CH_2 + 3\,CO + 2\,H_2O \longrightarrow R{-}CH_2{-}CH_2{-}CH_2OH + 2\,CO_2$$

The reaction has been carried to technical production by BASF [240, 1003]. It is employed on an industrial scale by Japan Butanol Company Ltd, Yokkaichi.

6.2. Dienes and Polyenes

Hydroformylation of conjugated dienes in the presence of $HCo(CO)_4$ gives only mono-aldehydes.

Dialdehydes have so far not been made. It is thought that hydroformylation of dienes gives α,β-unsaturated aldehydes, which are subsequently hydrogenated [243]. Heck and Breslow found a 1,4 addition of cobalt hydrocarbonyl to butadiene-1,3 at 0 °C [244] (1). Later work showed that, with cobalt deuterocarbonyl and butadiene-1,3, deuterium was added exclusively on the end carbon atom [245].

At higher temperature, a 2-butenoyl cobalt tricarbonyl complex is formed (2) [244, 246–249]. It is obvious that no dialdehyde can be formed from this allenyl compound. It probably reacts with hydrogen to form

Table 13. *Hydroformylation of conjugated, acyclic dienes and trienes*
(Cobalt catalyst)

Diene	Pressure (atm)	Temp. (°C)	Products	Yield (%)	Ref.
Butadiene	212–282	145–175	n- and iso-Valeraldehyde (1:1), dibutylketone	29	[192, 231]
2-Methyl-butadiene	212–282	150	Mixed C_6-aldehydes	16	[231]
1-Phenyl-butadiene	212	145–150	1-Phenylbutane (no hydroformyl-ation)	low	[231]
3-Methylpenta-diene-1,3			Mixed C_7-aldehydes	22	[231]
2,3-Dimethyl-butadiene			3,4-Dimethylpentanal	45	[231]
Cyclopentadiene	260	145–175	Cyclopentyl aldehyde	37	[231]
2,3-Dimethyl-butadiene	353	145–175	3,4-Dimethylpentanal 2,3-dimethylbutene-1 2,3-dimethylbutene-2	43	[231]
2,5-Dimethyl-hexadiene-2,4			Nonyl alcohol	high	[241]
Isoprene dimer	200	120	Acylic C_{11}-aldehydes and alcohols		[242]
Cyclohepta-triene			Hydroxymethyl-cyclo-heptane	45	[259]

butylene, which is then hydroformylated in the normal fashion to valeraldehyde. No experimental evidence exists as to whether it is possible to form unsaturated aldehydes through the reaction of carbon monoxide and hydrogen with the allenyl cobalt carbonyl complex.

$$H_2C=CH-CH=CH_2 + HCo(CO)_4 \xrightarrow{CO} H_3C-CH=CH-CH_2-\underset{\underset{O}{\|}}{C}-Co(CO)_4 \quad (1)$$

$$H_2C=CH-CH=CH_2 + HCo(CO)_4 \longrightarrow \underset{\substack{H}}{\overset{H_3C}{}}\!\!\underset{}{C}\!\!<\!\!\underset{\substack{C\\H\quad H}}{}\!\!Co(CO)_3 + CO \quad (2)$$

However, as reported by B. Fell and W. Rupilius [1034], dialdehydes can be made from conjugated diolefins using modified rhodium catalysts.

Thus, they reacted butadiene-1,3 and pentadiene-1,3 at 125 °C and 200 atm (cold) CO/H_2 (1 : 2) in diethyl ether in the presence of 0.02 mole-%

Rh$_2$O$_3$ and 1.7% tri-n-butylphosphine and obtained 80–90% yields of hydroformylation products which contained 40–45% of C$_6$- and C$_7$-dialdehydes respectively.

Surprisingly the C$_5$-monoaldehyde fraction contained more than 96% n-valeraldehyde and only 4% of iso-valeraldehyde.

The formation of the individual products may proceed via the following reaction sequence [1034]

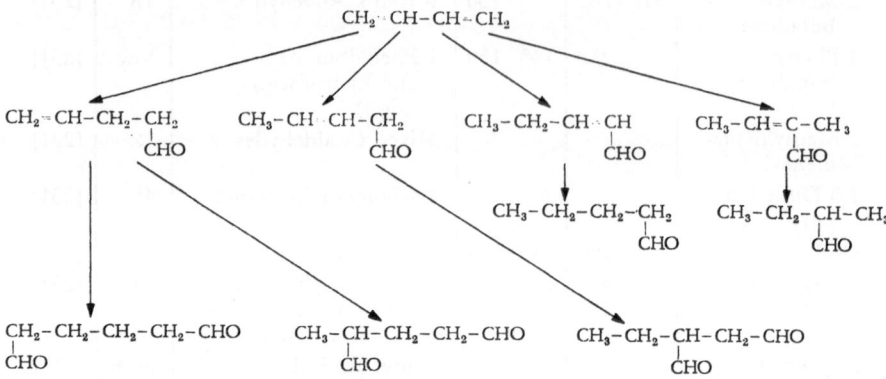

It may be assumed that dialdehyde formation can be achieved because the isomerisation of the intermediate nonconjugated unsaturated aldehyde to the conjugated aldehyde (which would only give monoaldehyde by hydrogenation) is not catalyzed by the modified rhodium catalyst (see also chapter on reaction mechanism of the hydroformylation reaction, p. 4), whereas HCo(CO)$_4$ catalyzes the isomerisation.

Dialdehydes can be obtained from dienes with isolated double bonds even with HCo(CO)$_4$. Yield of dialdehyde increases with increasing separation of the double bonds (table 14).

Table 14. *Hydroformylation of nonconjugated, alicyclic dienes*
Unless noted, cobalt catalyst was used

Diene	CO/H$_2$ Pressure (atm)	Temp. (°C)	Product	Yield (%)	Ref.
Pentadiene-1,4	1:1 [a]	150	60–100	C$_7$-Dialdehydes	19	[250]
Hexadiene-1,5	1:1 [a]	150	100	C$_8$-Dialdehydes		[250]
2,5-Dimethyl-hexadiene-1,5				nonanol, 3,6-Dimethyl-octanediol-1,8 (65 : 35)		[251]

[a] Rhodium catalyst

Cyclic diolefins with isolated double bonds give mixtures of mono- and dialdehydes (see table 15). The hydroformylation of these materials has been thoroughly investigated. Bis-(hydroxymethyl)-cycloaliphatics are very useful starting materials for polyester fibers and have been used for this purpose at the rate of several thousand tons per year [252].

The cyclooctadienes, easily obtained from butadiene-1.3, illustrated the complexity of the reaction paths of such dienes [253].

If the hydroformylation is carried out in the customary way with $Co_2(CO)_2$, at 110 °C for the hydroformylation stage and 200 °C for the hydrogenation stage and 200–300 atm ($CO/H_2 = 1:1$), a mixture of 63%. hydroxymethylcyclooctane and 26% bis-(hydroxymethyl)-cyclooctane is obtained.

The diol yield, with cobalt catalyst, is not increased with variation of the reaction conditions. Rhodium catalysts, on the other hand, give higher yields of diol under special reaction conditions. It is also possible to obtain unsaturated aldehyde, saturated aldehyde, or hydroxymethylcyclooctane as main product, depending on whether isomerization is enhanced or suppressed.

The two double bonds in (I) are not both hydroformylated at the same rate. At 80–100 °C and 100 atm CO/H_2 (1 : 1), only one of the double bonds reacts and cyclooctene aldehyde (IV) is formed. This can react further in two ways. At synthesis gas pressures above 200 atm, the second double bond is hydroformylated, at its original position, to a mixture of dialdehydes (V). Under 200 atm, the double bond is isomerized into conjugation with the aldehyde group. The α,β-unsaturated aldehyde (VI) is hydrogenated, slowly at 100 °C and rapidly at 150 °C, to the saturated aldehyde (VII), which may be hydrogenated to the monoöl (II) by raising the temperature to 210 °C.

Cyclooctadiene-1,5 is also partly isomerized during the reaction to cyclooctadiene-1,3 (VIII). Only the mono-aldehyde (VII) or monoöl (II) is obtained from (VIII), since conjugated dienes are hydrogenated rapidly to olefins (IX) in the hydroformylation reaction [10, 251]. (VIII) reacts practically quantitively through (IX) to (VII).

Table 15. *Hydroformylation of non-conjugated alicyclic dienes and trienes*

Unless otherwise noted, cobalt catalyst was used

Starting material	Pressure (atm)	Temp. (°C)	Products	Yield (%)	Ref.
Vinylcyclohexene	475–720	120–134	Mixed mono- and dialdehydes	65	[182]
Dicyclopentadiene[a]	180–195	155	Tricyclododecane dialdehyde	87	[254, 255, 991]
Cyclooctadiene-1,5	175	130	Hydroxymethyl-cyclooctane	70	[256]
			bis-(hydroxy-methyl)-cyclooctane	21.5	
Cyclooctadiene-1,5[a]	1000–1200	90/210	Hydroxymethyl-cyclooctane	13	[253]
			bis-(hydroxy-methyl)-cyclooctane	81	
Cyclododecatriene	150–300	135–145	Hydroxymethyl-cyclododecane	36.7 / 71[b]	[258] [968]
Bicycloheptadiene[a]	110/300	130/240	Bis-(hydroxy-methyl)-bicycloheptane	71	[969]

[a] Rhodium catalyst
[b] Phosphine modified Co-catalyst

Hydrogenation of both double bonds is very slight, cycloöctane being made in only trace amounts.

Dicyclopentadiene, which has no possibility of double bond isomerization, reacts smoothly to dialdehydes or, at higher temperature, to diols [254]. An isomer mixture is obtained in a ratio of 1 : 3.

If the reaction is carried out at 115 °C in alcohol solvents, the acetal of the dialdehyde is obtained. Partial hydrogenation of the dicyclopentadiene, and thereby formation of monoöl, was hardly observed.

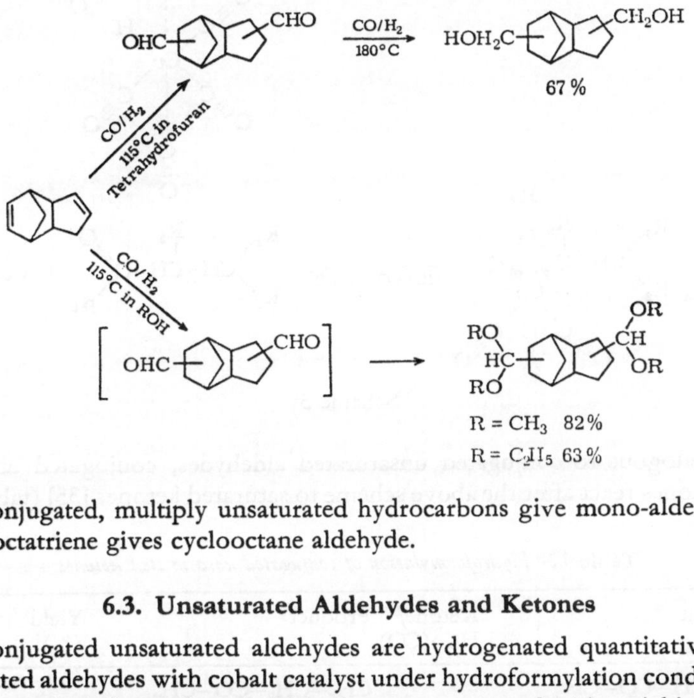

67 %

R = CH₃ 82%
R = C₂H₅ 63 %

Conjugated, multiply unsaturated hydrocarbons give mono-aldehydes: cyclooctatriene gives cyclooctane aldehyde.

6.3. Unsaturated Aldehydes and Ketones

Conjugated unsaturated aldehydes are hydrogenated quantitatively to saturated aldehydes with cobalt catalyst under hydroformylation conditions.

A number of unsaturated aldehydes were contacted with stoichiometric amounts of cobalt hydrocarbonyl at normal conditions by Orchin *et al.* [260], (see table 16).

Table 16 [260]. *Reaction of unsaturated aldehydes with cobalt hydrocarbonyl*
25 °C, 1 atm

Reactant	Product	Yield (%)
Acrolein	Propionaldehyde	93
Crotonaldehyde	Butyraldehyde	80
α-Methylcrotonaldehyde	α-Methylbutyraldehyde	15
Cinnamaldehyde	3-Phenylpropionaldehyde	97

In these cases, the reaction is exclusively hydrogenation. The reason for the absence of hydroformylation lies, according to Orchin *et al.*, in the formation of a π-allyl type complex intermediate, which reacts according to scheme 3 (see also the section on dienes).

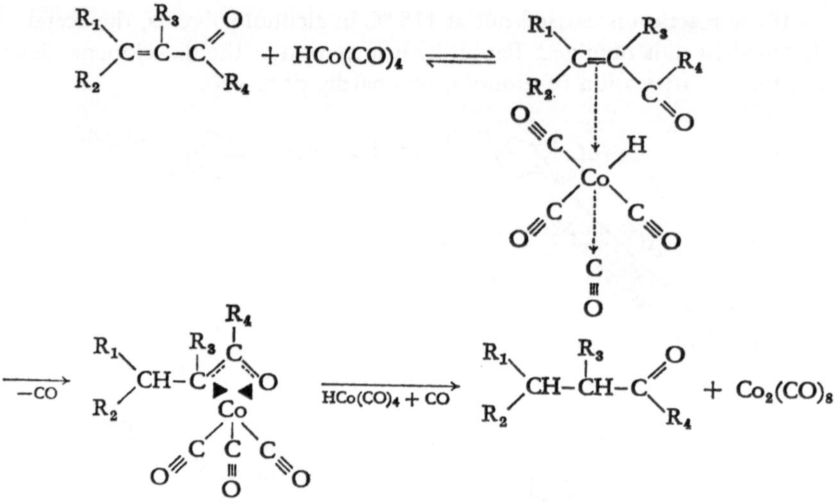

Scheme 3

Analogous to conjugated unsaturated aldehydes, conjugated unsaturated ketones react after the above scheme to saturated ketones [35] (table 17).

Table 17. *Hydroformylation of conjugated unsaturated ketones*

Reactant	Ketone/ HCo(CO)$_4$	Product	Yield (%)	Ref.
$CH_2=CH-CO-CH_3$		$CH_3-CH_2-CO-CH_3$		[24]
$(CH_3)_2C=CH-CO-CH_3$		$(CH_3)_2CH-CH_2-CO-CH_3$		[24]
$CH_2=CH-CO-CH_3$	1:1	$CH_3-CH_2-CO-CH_3$	70	[260]
$C_6H_5CH=CH-CO-CH_3$	1:1	$C_6H_5-CH_2-CH_2-CO-CH_3$	56	[260]
$(CH_3)_2-C=CH-CO-CH_3$	1:1	$(CH_3)_2-CH-CH_2-COCH_3$	24	[260]

Conjugated unsaturated aldehydes and ketones may be hydroformylated in good yield through their acetals or ketals (see also the section on unsaturated ethers).

When the double bond in unsaturated aldehydes is not in conjugation with the carbonyl group, dialdehydes are obtained in good yield. Thus, tetrahydrobenzaldehyde gives dialdehyde, or diol at higher temperature, in good yield [174].

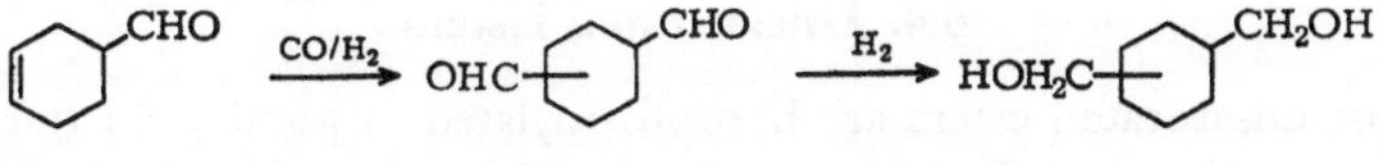

The carbonyl group in a molecule has an important directing effect on the location of the hydrocarbonyl addition. Formation of 1,3-dialdehydes is favored. Especially high yields, with a higher proportion of 1,4-dialdehyde or subsequent diol, are obtained with rhodium catalyst [174].

The 6:4 mixture of endo- and exo-bicycloheptenealdehyde [263], obtained from the Diels-Alder reaction [262] of acrolein with cyclopentadiene, was hydroformylated with rhodium catalyst [254].

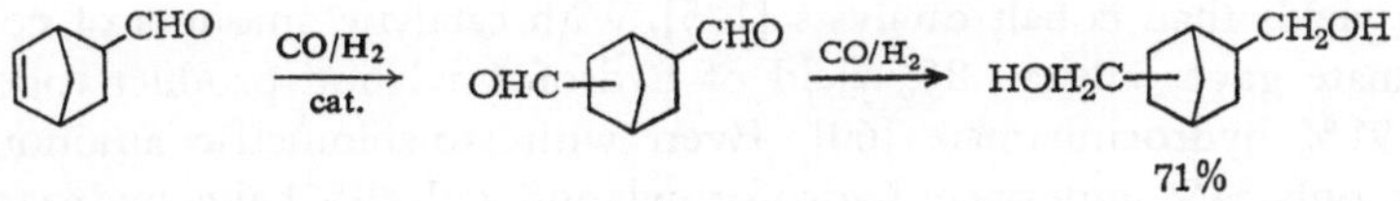

71%

Theoretically, six isomeric dialdehydes, or diols at higher temperature, are possible:

exo,exo-2,6-Bis-(hydroxymethyl)-bicyclo-2,2,1-heptane (a),
exo,exo-2,5-Bis-(hydroxymethyl)-bicyclo-2,2,1-heptane (b),
exo,endo-2,6-Bis-(hydroxymethyl)-bicyclo-2,2,1-heptane (c),
exo,endo-2,5-Bis-(hydroxymethyl)-bicyclo-2,2,1-heptane (d),
endo,endo-2,6-Bis-(hydroxymethyl)-bicyclo-2,2,1-heptane (e),
endo,endo-2,5-Bis-(hydroxymethyl)-bicyclo-2,2,1-heptane (f),

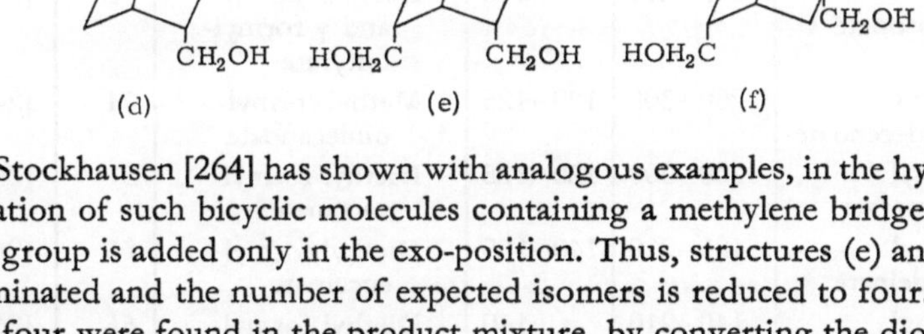

(a)   (b)   (c)

(d)   (e)   (f)

As Stockhausen [264] has shown with analogous examples, in the hydroformylation of such bicyclic molecules containing a methylene bridge, the formyl group is added only in the exo-position. Thus, structures (e) and (f) are eliminated and the number of expected isomers is reduced to four.

All four were found in the product mixture, by converting the diols to their silyl ethers and separating them, by gas chromatography, in the ratios 21 : 27 : 34 : 18.

If the two bicycloheptenealdehydes are separated and hydroformylated separately, two isomer products are obtained from each.

6.4. Unsaturated Esters

Most unsaturated esters are hydroformylated in good yield (table 18). The conjugation in α,β-unsaturated esters is weaker than in unsaturated aldehydes (the resonance energy of crotonaldehyde, for example, is 2.4 Kcal/mole higher than for ethyl crotonate [7]). Thus, conjugated unsaturated esters such as acrylates and crotonates, in contrast to acrolein and crotonaldehyde, can be converted to aldehyde products with synthesis gas.

Even unsaturated esters having the double bond in conjugation with another unsaturated system, such as maleinates, fumarates and cinnamates, are hydroformylated (see table 18). Rhodium catalysts frequently give better yields than cobalt catalysts [175]. With catalytic amounts of cobalt, cinnamate gave only an 8% yield of hydroformylation product together with 91% hydrocinnamate [60]. Even with stoichiometric amounts of cobalt, only 34% underwent hydroformylation with 49% being hydrogenated to the hydrocinnamate. In contrast, with catalytic amounts of rhodium, the yield of hydroformylation product was 73% [175].

Table 18. *Hydroformylation of unsaturated esters*
$$CO/H_2 = 1:1$$

Reactant	Pressure (atm)	Temp. (°C)	Products	Yield (%)	Reference
Methyl acrylate	200	120	Methyl β-formyl-propionate	85	[265]
Ethyl acrylate	200–300	120–125	Ethyl α- and β-formyl-propionate	75	[24, 266]
Ethyl crotonate	200–300	120–125	Ethyl α- and β-, and γ-formyl-butyrate	71	[24, 266]
Butyl crotonate	145–210	140	Butyl α- and β- and γ-formyl-butyrate	78	[267]
Methyl undecenoate	200–300	120–125	Methyl formyl-undecanoate	71	[24]
Diethyl fumarate	200–300	120–125	Diethyl formyl-succinate	51	[268]
Diethyl maleinate	183–218	140–155	Diethyl formyl-succinate	65	[267]
Diethyl itaconate	140–210	140	Diethyl formyl-methylsuccinate	56	[267]
Methyl oleate	600–750[a]	140–145	Mixed aldehyde esters	72	[269–271]

[a] $CO/H_2 = 1:2$

In analogy with olefins, the hydroformylation of unsymmetrical unsaturated esters gives a mixture of isomeric products. α,β-Unsaturated esters differ from olefins in that the α-carbon atom carries the highest negative charge. At low temperatures and high pressures, the α-formyl compound is formed, while the β-formyl ester is produced at higher temperatures and lower pressures [59, 60].

$$\text{H}_2\text{C=CH--COOR} + \text{CO/H}_2 \begin{cases} \text{H}_3\text{C--CH--COOR} \\ \quad\quad\quad | \\ \quad\quad\quad \text{CHO} \\ \\ \text{OHC--CH}_2\text{--CH}_2\text{--COOR} \end{cases}$$

Experiments with stoichiometric amounts of cobalt hydrocarbonyl and acrylate ester at 0 °C gave a product with $\alpha : \beta$-formyl ratio of 5 : 1 [35].

Fig. 9 shows the strong dependence of the isomer composition on the reaction temperature [60].

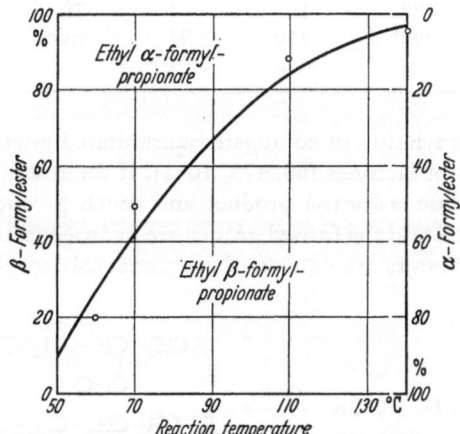

Fig. 9. Distribution of α- and β-isomers of ethyl formylpropionate formed by the hydroformylation of acrylates as a function of reaction temperature. 300 atm (CO/H$_2$ = 1:1); toluene solvent

Long chain unsaturated esters are susceptible to isomerization, by a scheme similar to that given on page 12 for olefins. Thus, from crotonate esters, a larger amount of γ-formylbutyrate is formed, besides, α- and β-formylbutyrate [59–61, 266]. Rhodium catalysts give a higher yield of α-formyl product at otherwise similar reaction conditions [175]. Thus, with methacrylate esters, which gave only the β-formyl product with cobalt at higher temperatures, an 80% yield of α-formylisobutyrate can be made (see table 19).

$$H_2C=\underset{\underset{CH_3}{|}}{C}-COOR + CO/H_2 \xrightarrow{Rh} H_3C-\underset{\underset{CHO}{|}}{\overset{\overset{CH_3}{|}}{C}}-COOR$$

$$OHC-CH_2-\underset{\underset{CH_3}{|}}{CH}-COOR$$

In this case, contrary to the rules of Keulemans, a compound containing a quarternary carbon atom is made in high yield.

Table 19. *Hydroformylation of methyl methacrylate with various catalysts*

Catalyst	Press. (atm)	Temp. (°C)	α-Formyl product (%)	β-Formyl product (%)	Iso-butyrate (%)	Reference
Cobalt	300	140	—	51	42	[60]
Rhodium	200	165	4	78	10	[175]
Rhodium/tributyl phosphine	600	110	94	trace	trace	[175]

The hydroformylation of conjugated unsaturated esters can also lead to a direct synthesis of lactones [60, 175, 1041]. If the reaction conditions are chosen so that little α-formyl product and much β- and γ-formyl compounds are formed and the formyl groups are hydrogenated to alcohols, the resulting hydroxyesters are converted spontaneously to γ- or δ-lactones in high yield.

$$RCH_2-CH=CH-COOR \xrightarrow{CO/H_2}$$

$$RCH_2-\underset{\underset{CHO}{|}}{CH}-CH_2-COOR \xrightarrow{H_2}$$

$$R-\underset{\underset{CHO}{|}}{CH}-CH_2-CH_2-COOR \xrightarrow{H_2}$$

$$RCH_2-\underset{\underset{CH_2OH}{|}}{CH}-CH_2-COOR \xrightarrow{-ROH}$$

$$R-\underset{\underset{CH_2OH}{|}}{CH}-CH_2-CH_2-COOR \xrightarrow{-ROH}$$

In analogy to the hydroformylation of dienes, the doubly unsaturated sorbate ester gave no diformyl product but only the mono-formyl compound [60] (see also page 40).

Even the modified rhodium catalyst, which produces dialdehydes from conjugated dienes [1034] (see also chapter on hydroformylation of dienes) yields only the monohydroformylated product.

Esters made from unsaturated alcohols usually react smoothly with synthesis gas: vinyl acetate gave 30% α-acetoxypropionaldehyde together with 22% of the β-isomer [24], and allyl acetate produced α-acetoxy-butyraldehyde in 75% yield [24]. Under certain conditions, it is possible to split out the acid group and obtain the unsaturated aldehyde [272].

Table 20 [60, 175]. *γ- and δ-Lactones from unsaturated esters*

Unless otherwise noted, $Co_2(CO)_8$ was used as catalyst

Reactant	Product	Yield (%)
Methyl acrylate	γ-Butyrolactone	69
Ethyl acrylate	γ-Butyrolactone	88
Methyl methacrylate	α-Methyl-γ-butyrolactone	52
Methyl methacrylate	α-Methyl-γ-butyrolactone	78[a]
Ethyl cyclohexenylformate	2,3-Tetramethylene-γ-butyrolactone	23
Methyl crotonate	δ-Valerolactone +	72
	β-methyl-γ-butyrolactone	20
Ethyl crotonate	δ-Valerolactone +	67
	β-methyl-γ-butyrolactone	23
Ethyl tiglinate	α-Methyl-δ-valerolactone +	31
	α-ethyl-γ-butyrolactone +	21
	α,β-dimethyl-γ-butyrolactone	
Ethyl vinylacetate	δ-Valerolactone +	52
	β-methyl-γ-butyrolactone	17
Ethyl β,β-dimethylacrylate	β-Methyl-δ-valerolactone +	88
	β,β-dimethyl-γ-butyrolactone	1
Ethyl α,β,β-trimethyl-acrylate	α,β-Dimethyl-δ-valerolactone +	55
	α-isopropyl-γ-butyrolactone	31
Ethyl dimethylvinylacetate	α,α-Dimethyl-δ-valerolactone	93
	α,α,β-trimethyl-γ-butyrolactone	1
Ethyl sorbate	β-Propyl-γ-butyrolactone +	33
	γ-ethyl-δ-valerolactone	49
Ethyl cinnamate	β-Phenyl-γ-butyrolactone	8.5
Ethyl cinnamate (stoic. amt. Co cat.)	β-Phenyl-γ-butyrolactone	34
Ethyl cinnamate	β-Phenyl-γ-butyrolactone	73[a]
Diethyl fumarate	β-Carbethoxy-γ-butyrolactone	49
Diethyl maleinate	β-Carbethoxy-γ-butyrolactone	47

[a] Rhodium catalyst

$$H_2C=CH-O-\underset{\underset{O}{\|}}{C}-R + CO/H_2 \longrightarrow H_3C-\underset{CHO}{\overset{|}{C}}H-O-\underset{\underset{O}{\|}}{C}-R$$

$$\longrightarrow H_2C=CH-CHO + RCOOH$$

6.5. Unsaturated Nitriles

Conjugated and unconjugated unsaturated nitriles are both susceptible to hydroformylation. Few examples are found in the literature and most of the work has been with acrylonitrile (table 21). The aldehyde group is incorporated mainly in the β-position, but the reaction is strongly solvent dependent and byproducts are frequently produced, mainly propionitrile, acrolein, propyl amine and ammonia (see also ref. [972, 973] and [985]).

Table 21. *Hydroformylation of acrylonitrile*

CO/H₂	Press. (atm)	Temp. (°C)	Solvent	Product	Yield (%)	Ref.
	900	150	CH₃OH	β-Cyanopropion-aldehyde-diacetal	75	[273, 274]
1:1	200	120–130	(CH₃)₂CO	OHC–CH₂–CH₂CN byproducts:	82	[274, 275, 786]
			1 mole % Co₂(CO)₂	CH₃–CH₂–CN		
				CH₃–CH₂–CHO HO(CH₂)₃–CN C₃H₇–NH₂, NH₃		
1:1	550	100		OHC–CH₂–CH₂–CN	86	[276]
1:1	220	70–100	Paraffins Aromatics Alcohols Glycols	OHC–CH₂–CH₂–CN	57	[277]

Allyl cyanide gives γ-cyanobutyraldehyde in 36% yield [220, 269, 271]. 5-Cyano-2-methylpentene-1, at 150–170 °C and 700 atm CO/H_2 (1:1), gave 45% yield of 6-cyano-3-methylhexanal-1 with a trace of 5-cyano-2,2-dimethylpentanal-1, while at 239–260 °C and the same pressure, 36% of 6-cyano-3-methyl-hexanal-1 was made [278].

6.6. Unsaturated Alcohols

Unsaturated alcohols of the allyl alcohol type are hydroformylated poorly [24, 29, 205, 279]. Allyl alcohol reacts to give γ-hydroxybutyralde-hyde and β-hydroxyisobutyralehyde:

$$HO-CH_2-CH=CH_2 + CO/H_2 \longrightarrow HO-CH_2-CH_2-CH_2-CHO$$

$$HO-CH_2-CH-CHO$$
$$\underset{CH_3}{|}$$

The yields do not rise over 30%, due to the formation of a number of byproducts. Especially obstructive is the isomerization of the starting material to propionaldehyde [280].

$$HO-CH_2-CH=CH_2 \longrightarrow HO-CH=CH-CH_3 \longrightarrow OHC-CH_2-CH_3$$

Unsaturated alcohols which have the double bond farther removed from the hydroxyl group are reacted in good yields. The reaction of tetrahydro-benzyl alcohol was investigated in detail [174].

A mixture of cis- and trans-1,3- and 1,4-bis-(hydroxymethyl)-cyclo-hexane (52:48) in 70% yield was obtained in methanol at 180 °C. 11% of the starting material was hydrogenated to cyclohexyl carbinol.

If the reaction is carried out under 150 °C, the isomers can be separated, since the cis-compounds react to cyclic acetals, while the trans isomers remain as the hydroxyaldehydes. At higher temperatures, the acetal group can be split off.

This hydroformylation of unsaturated alcohols having strongly steri-cally shielded double bonds has been described by D. C. Hull *et al.* [281]. As expected, the formyl group is attached to carbon atom 5 (see the details over olefins on page 38).

An unsaturated alcohol where the double bond is conjugated to an aro-matic ring was hydroformylated by Nahum [970]. From coniferyl alcohol he obtained a 25% yield of 3-(3-methoxy-4-hydroxyphenyl)-tetrahydro-furan the formation of which is explained by ring closure of the primary hydroformylation product. Byproducts were: 3(3-methoxy-4-hydroxy-phenyl)propanol, 4-n-propylguaiacol, 3-methoxy-5,6,7,8-tetrahydro-2-naph-thol, 4-ethyl guaiacol, guaiacol, phenol and probably 2-(3-methoxy-4-hydro-xyphenyl)tetrahydrofuran.

6.7. Unsaturated Ethers and Acetals

The hydroformylation of ethers has been known for a long time, but only recently has it been investigated systematically.

In vinyl ether, the formyl group goes on the α-carbon atom, under standard hydroformylation conditions (125 °C/200–300 atm). Butyl vinyl ether gives α-n-butoxypropionaldehyde in 31 % yield [24].

$$H_2C=CH-O-C_4H_9 \xrightarrow{CO/H_2} H_3C-CH-O-C_4H_9$$
$$\overset{|}{CHO}$$

It is not yet clear whether the electron withdrawing influence of the ether oxygen is noticeable even across a methylene group. Ethyl allyl ether [24] e.g. was reported to give 30% β-ethoxyisobutyraldehyde with only 4% γ-ethoxy-n-butyraldehyde. The methacrolein isolated (6%) was probably formed by loss of alcohol from the formed β-ethoxy-n-butyraldehyde.

On the other hand allyl phenyl ether was found to give only phenoxybutyraldehyde with a selectivity up to 84% [971].

In analogy with the results of unsaturated esters, unbranched aldehydes can also be obtained from vinyl ethers at higher temperatures. Thus,

Gresham and Brooks isolated β-methoxypropionaldehyde after the hydroformylation of methyl vinyl ether at 160 to 175 °C [282]. A series of easily accessible dihydropyrans were also studied [283]. Also in dihydropyran, the formyl group prefers to go on carbon atom 2 [283–286].

If the 2-position is alkyl substituted, the formyl group goes almost exclusively on carbon atom 3.

Methyl 6-methyl-5,6-dihydro-4H-pyrancarboxylate-2 reacts in a similar manner at higher temperatures, although a part of the starting material was hydrogenated.

$$H_3C\text{-}O\text{-}COOCH_3 \xrightarrow[\text{Co}_2(\text{CO})_8]{\text{CO/H}_2} \left[H_3C\text{-}O\text{-}COOCH_3\text{-}CH_2OH \right] \longrightarrow$$

$$H_3C\text{-}O\cdots O + CH_3OH$$

Unsaturated acetals react very much like unsaturated ethers [287, 288],

$$H_3C\text{-}CH\text{=}CH\text{-}\underset{\underset{OC_2H_5}{|}}{\overset{\overset{OC_2H_5}{|}}{CH}} \xrightarrow{\text{CO/H}_2} H_3C\text{-}CH\text{-}CH_2\text{-}\underset{\underset{OC_2H_5}{|}}{\overset{\overset{OC_2H_5}{|}}{CH}}$$
$$\qquad\qquad\qquad\qquad\qquad\qquad\qquad CHO$$

$$H_2C\text{=}CH\text{-}C\underset{O\text{-}CH\text{-}CH_3}{\overset{O\text{-}CH\text{-}CH_3}{<}} \xrightarrow{\text{CO/H}_2} H_3C\text{-}CH\text{-}C\underset{O\text{-}CH\text{-}CH_3}{\overset{O\text{-}CH\text{-}CH_3}{<}} \qquad [205]$$
$$\qquad\qquad\qquad\qquad\qquad\qquad\qquad\qquad CHO$$

thus giving the possibility of carrying out the hydroformylation reaction on unsaturated aldehydes and ketones which cannot be directly converted (see also the section on unsaturated aldehydes and ketones).

6.8. Unsaturated Halogen Compounds

For a long time, it was considered that unsaturated halogen compounds could not be hydroformylated. It was reported that the cobalt catalyst would be halogenated and also that the dehalogenated product could not be hydroformylated.

Table 22. *Hydroformylation of unsaturated halogen compounds*

Reactant	Product	Yield (%)	Ref.
P-Cl-C$_6$H$_4$OCH$_2$CH=CH$_2$	Aldehydes	11	[289]
Hexafluoropropylene	Hexafluoropropane, alcohols, aldehydes	50/40 5–8	[290]
Vinyl chloride	α-Cl-propionaldehyde, acrolein, propionaldehyde, ethyl chloride	85–90	[291, 292]
1,1-Dichloroethylene [a]	Dichloroethane, α-Cl-propionaldehyde, α-Cl-acrolein		[291]
1-Bromopropene	α-Br-butyraldehyde		[292]

[a] Reaction with stoichiometric amount of HCo(CO)$_4$ at −15 °C

Recently, some halogenated materials have been found which can be reacted with synthesis gas. These are compounds which have the halogen relatively tightly bound: chlorinated aromatics [289], perfluoro olefins [290] or vinylic chloride compounds [291, 292] (table 22).

6.9. Acetylenes

Acetylenes are much less adapted to hydroformylation than olefins. Very few examples are reported in the literature. Roelen reported [23] that acetylene reacted with synthesis gas in the presence of cobalt even at low pressure (10 atm, 140–150 °C); the primary reaction was the formation of acrolein.

Greenfield, *et al.* [293] were able to detect propionaldehyde from the reaction of acetylene with cobalt hydrocarbonyl at room temperature. In the reaction of pentyne-1 with synthesis gas, they obtained 6% n-hexanol and 5.5% 2-methylpentanol-1. Although many attempts were made [24], phenyl acetylene could not be hydroformylated. Also, diphenyl acetylene was only hydrogenated and gave 1,2-diphenylethane and cis-stilbene. The hydroformylation of acetylene also goes with rhodium catalyst, but only with moderate yields. Hexyne-1, contacted with $(Ph_3P)RhCl_2$ in ethanol/benzene at 110 °C and 120 atm $H_2/CO = 1 : 4$, gave a total yield of 15% of an equimolar mixture of n-heptanol and 2-methylhexanol [294].

6.10. Aromatics and Heterocyclics

The double bonds in aromatic hydrocarbons cannot be hydroformylated. Thus, aromatics like benzene, toluene, and xylene can be used as solvents for the hydroformylation reaction even under extreme conditions without themselves being reacted.

Heterocyclics with aromatic character are intermediates between dienes and aromatic hydrocarbons. Under drastic conditions, thiophene and its

derivatives are hydrogenated to tetrahydrothiophenes [295], and pyridines go to piperidines. The formed piperidines react with carbon monoxide to N-formylpiperidines [295], which are partly ring opened and a large number

of products are formed. Furan and 2,5-dimethylfuran can be hydroformylated to give 2-tetrahydrofurfuryl alcohol and 2,5-dimethyl-3-tetra-
hydrofurfuryl alcohol, respectively.

They behave like dienes in the hydroformylation reaction. The
addition follows the same steps as already shown for dihydropyrans in
the section on unsaturated ethers.

6.11. Epoxides

Many examples of the reaction of epoxides with synthesis gas are described in the literature [34, 35, 225, 296–298]. Heck was able to show, by
stoichiometric reactions at normal conditions, that this reaction also proceeds through acyl cobalt carbonyls [301].

$$H_2C\overset{O}{\diagup\!\!\!\diagdown}CH_2 + HCo(CO)_4 \longrightarrow HOCH_2-CH_2-Co(CO)_4$$
$$HOCH_2-CH_2-Co(CO)_4 + CO \longrightarrow HOCH_2-CH_2-CO-Co(CO)_4$$

The acyl cobalt carbonyls are very probably then hydrogenated to
aldehydes and cobalt hydrocarbonyl, in analogy to the described mechanism
for the olefins. As Yokokawa, *et al.* [298] found, the yield depends strongly
on the reaction temperature. At higher temperatures (see table 23), isomerization of the epoxides to the corresponding ketones and aldehydes
frequently occurs first. Further reaction gives a large number of polymeric products.

Table 23 [302]. *Hydroformylation of propylene oxide*
$Co_2(CO)_8$, $CO/H_2 = 1:1$, 150–160 atm

Product composition (moles/100 moles C_3H_6O)	90–95 °C	115–120 °C
Acetone	6.4	51.0
Isobutyraldehyde	1.2	5.1
Methacrolein	0.8	Traces
n-Butyraldehyde	1.4	16.4
Isobutanol	Traces	1.5
Crotonaldehyde	5.4	Traces
n-Butanol	Traces	2.3
β-Hydroxy-isobutyraldehyde	—	Traces
β-Hydroxy-n-butyraldehyde	56.0	3.3

The main reaction products are usually the β-hydroxyaldehydes. Under more severe conditions, unsaturated aldehydes are formed through dehydration, which can be hydrogenated to saturated aldehydes under the hydroformylation conditions.

The hydroformylation of ethylene oxide gives acrolein in very low yields, which is not surprising in view of its extraordinary reactivity; in the main, resinous products are obtained. The reactions of ethylene oxide, propylene oxide, cyclohexene oxide, styrene oxide and epichlorhydrin were studied by Takegami et al. [297] at normal conditions. They found that generally internal and conjugated olefin oxides were more reactive than terminal ones and established the following order of reactivity: cyclohexene oxide (appr. 5) > styrene oxide > propylene oxide (1) > ethylene oxide ≫ epichlorhydrin (1/20–1/10) (the numbers give the relative reactivity).

From the work of Takegami et al., and Heck, it must follow that cobalt hydrocarbonyl reacts with epoxides in the acid form. The reaction of epoxide compounds should be accelerated by addition of small amounts of alcohols, ketones, ethers and esters [303], as well as copper oxide or halide, silver oxide and aluminium chloride [304].

Orchin et al. [305] recently showed that the reaction of cyclohexene oxide, at normal conditions with catalytic amounts of cobalt hydrocarbonyl, gave its dimer, a cyclic hemiacetal, instead of the monomeric hexahydrosalicylaldehyde.

6.12. Saturated Alcohols (Homologation)

Saturated alcohols can also be hydroformylated, giving aldehydes or alcohols larger by one methylene group [306, 307]. The reaction has also been called "Homologation" [308]. The reaction rate of the alcohols decreases in the order: tertiary, secondary, primary. Benzyl alcohol is very reactive.

Also, acetals of aldehydes react in this fashion. Thus, a mixture of glycol monomethyl ether and methoxyacetaldehyde dimethylacetal is obtained from formaldehyde dimethylacetal [309, 310].

There are many indications that in the homologation reaction the cobalt hydrocarbonyl reacts in its acid (ionized) form and a carbonium ion is formed from the alcohol by lose of the -OH group. This point of view has

Table 24 [311]. *Hydroformylation of alcohols*

Alcohol	Products	Reference
Methanol	Acetaldehyde, propionaldehyde, methyl acetate	[312]
	39 % Ethanol, 5 % propanol, 1 % butanol, 9 % Methyl acetate, 6.3 % Ethyl acetate, 8.5 % Methane	[227, 314]
Ethanol	n- and iso-Butanol, n- and iso-Pentanol	[308, 314]
Propanol-1	11 % n- and iso-Butanol	[308]
tert.-Butanol	51 % Isovaleraldehyde, 10 % pivaldehyde 60 % Isoamyl alcohol, 4 % Neopentyl alcohol 26 % High-boiling material	[230, 308, 314–316]
Cyclohexanol	44 % Cyclohexyl carbinol	[314]
Pinacol	26 % 3,4-Dimethylpentanol 17 % Pinacolone, 4 % pinacolyl alcohol	[317]
Benzyl alcohol	32 % 2-Phenylethanol, 63 % toluene	[318]
p-Methylbenzyl alcohol	24 % 2-(p-Methylphenyl)-ethanol, 58 % p-Xylene	[318]
m-Methylbenzyl alcohol	36 % 2-(m-Methylphenyl)-ethanol, 52 % m-Xylene	[318]
p-tert-Butylbenzyl alcohol	28 % 2-(p-tert-Butylphenyl)-ethanol 54 % p-tert.-Butyltoluene	[318]
2,4,6-Trimethyl-benzyl alcohol	18 % 2-(2,4,6-Trimethylphenyl)-ethanol 58 % 1,2,3,5-Tetramethylbenzene	[318]
p-Hydroxymethyl-benzyl alcohol	39 % 2-(p-Methylphenyl)-ethanol 12 % p,p'-Phenylene-β,β'-diethanol 27 % p-Xylene	[318]
p-Methoxy-benzyl alcohol	44 % 2-(p-Methoxyphenyl)-ethanol 16 % p-Methoxy-toluene	[318]
m-Methoxy-benzyl alcohol	2 % 2-(m-Methoxyphenyl)-ethanol 23 % m-Methoxy-toluene	[318]
p-Chlorobenzyl-alcohol	16 % 2-(p-Chlorophenyl)-ethanol 41 % p-Chloro-toluene	[318]
m-Trifluoromethyl-benzyl alcohol	5 % m-Methylbenzotrifluoride	[318]
p-Carbethoxybenzyl alcohol	27 % Ethyl p-methylbenzoate	[318]
p-Nitrobenzyl alcohol	Polymeric p-aminobenzyl alcohol	[318]

been supported by the investigations of Wender, *et al.* [318], who studied the reaction rates of various substituted benzyl alcohols and showed that, if the substituents were in a position to stabilize the carbonium ion through resonance, the reaction rate of the benzyl alcohol was increased substantially.

At the same time the yields of the homologation alcohols increased while the yield of hydrogenated product, usually obtained as byproduct, decreased.

Certain aromatic ketones can also be homologated. Thus, p-methoxy-acetophenone reacts to 2-(p-methoxyphenyl)-propanol-1. It must be assumed that the ketone is first hydrogenated to the alcohol, which then undergoes homologation.

$$H_3CO-\langle\text{phenyl}\rangle-\underset{\underset{O}{\|}}{C}-CH_3 \longrightarrow H_3CO-\langle\text{phenyl}\rangle-CHOH-CH_3 \longrightarrow$$

$$H_3CO-\langle\text{phenyl}\rangle-\underset{\underset{CH_2OH}{|}}{CH}-CH_3$$

Table 25 [10]. *Relative reaction rates for substituted benzyl alcohols*

Substituent	Time[a], min	Temp.[b], °C	Relative rate[c]
p-Methoxy	6	92	1×10^4
p-Methyl	44	166	2×10^2
m-Methyl	67	188	5×10
p-tert-Butyl	67	188	5×10
Hydrogen	82	190	1
p-Chloro	109	190	0.8
p-Carbethoxy[d]	231	190	0.4
m-Methoxy[d]	254	190	0.3
m-Trifluoromethyl	~10,000[e]	190	0.01

[a] Time for absorption of 1 mole gas/mole alcohol at 80 °C
[b] Temperature after the time interval of column 2. The last four alcohols began to react at 190 °C. Benzyl alcohol took up gas slowly at 180 °C
[c] Under the assumption that the rate doubles for every 10 °C increase in temperature. Otherwise, this column gives the ratio of absorption times (1 mole gas/mole alcohol) of benzyl alcohol to the substituted alcohol, both at 190 °C
[d] The rate data for p-carbethoxy and m-methoxy are for 0.35 mole sample and therefore cannot be compared exactly with the rest of the data using 0.6 mole samples
[e] Estimated, only 5% of the alcohol reacted in 5 hours

Table 26 [10]. *Reaction of substituted benzyl alcohols with synthesis gas*
$CO/H_2 = 1 : 2$; initial pressure 245 atm, maximum reaction time:
5 hours at 185–190 °C

Substituent	Tolu-ene product (A), %	Ethanol product (B), %	% B/A	Recov-ered reac-tant, %	Other products
p-Methoxy	16	44	2.8	0	appr. 34 % High-boiling polymer, probably aldol from p-methoxyphenyl acetaldehyde
p-Hydroxymethyl	27	39, 12[b]	1.4	0	
p-Methyl	58	24	0.4	0	8 % Polymer, probably aldol[a]
m-Methyl	52	36	0.7	0	1–2 % m-Methyl-benzaldehyde
p-tert-Butyl	54	28	0.5	0	1 % p-tert. Butyl-benzaldehyde
Benzyl	63	31	0.5	0	
2,4,6-Trimethyl	58	18	0.3	0	
p-Chloro	41	16	0.4	31	4 % p-Chlorobenz-aldehyde
m-Methoxy	23	appr. 2	0.1	56	4% m-Methoxybenz-aldehyde
m-Trifluoro-methyl	5			78	4% m-Trifluoro-methylbenz-aldehyde
p-Carbethoxy	27				High-boiling products
p-Nitro				44	Polymers of p-amino-benzyl alcohol

[a] This product was added to the yield of alcohol in column 3
[b] This yield is for the dihomologized alcohol, p,p′-phenylene-β,β′-diethanol

7. Aldol Condensation During Hydroformylation

Frequently in the hydroformylation of olefins to aldehydes, a further reaction in varying extent (usually undesired) is aldol condensation.

There are certain cases, however, where aldol condensation is desired. For instance, by the aldol condensation of butyraldehyde, followed by hydrogenation, one obtains 2-ethylhexanol [893, 894, 952], a very important plasticizer alcohol, produced in the hundreds of thousands of tons per year. This compound and its homologs can be produced by hydroformylation in a one-step or two-step process if a catalyst for aldol condensation is included with the hydroformylation catalyst. In this connection, basic substances are of interest. Table 27 shows how the inclusion of $Mg(OCH_3)_2$ in the hydroformylation reaction can increase the amount of dimeric product.

Table 27 [319]. *Formation of 2-ethylhexanal by the hydroformylation of propylene*
$$CO/H_2 = 1:1; \ 200 \ atm; \ 150 \ °C$$

Catalyst	Reaction time (min)		
	20	40	60
$Co_2(Co)_8$	10%[a]	17%	22%
$Co_2(CO)_8 + Mg(OCH_3)_2$	43%	53%	60%

[a] wt-% C_8 aldehyde in product

Table 27 shows the influence of magnesium methylate/pyridine and iron pentacarbonyl on the formation of 2-ethylhexanal in the hydroformylation of propylene with cobalt catalyst.

$$H_3C-CH=CH_2 + CO/H_2 \xrightarrow[\text{hydrof.-cat.}]{150\,°C} H_3C-CH_2-CH_2-CHO \qquad (1)$$

$$2\,H_3C-CH_2-CH_2-CHO \xrightarrow[\text{cond.-cat.}]{150\,°C} H_3C-CH_2-CH_2-CH=\underset{\underset{C_2H_5}{|}}{C}-CHO + H_2O \qquad (2)$$

$$H_3C-CH_2-CH_2-CH=\underset{\underset{C_2H_5}{|}}{C}-CHO + H_2 \xrightarrow[\text{hydroformyl.-cat.}]{200\,°C}$$
$$H_3C-CH_2-CH_2-CH_2-\underset{\underset{C_2H_5}{|}}{CH}-CH_2OH \qquad (3)$$

In this way, more than two thirds of the formed n-butyraldehyde can be converted directly to 2-ethylhexanol.

The described reactions have been carried to technical production (Esso) under the name "Aldox" reaction, utilizing a condensation catalyst. Compounds of Zn, Sn, Ti, Zr, Hf, Th, Pb, Cd, Hg, Al and Cu may also be used for this purpose [709, 715–717, 757, 758, 782–785]. Shell also produces these alcohols in a similar reaction and uses KOH as a condensation catalyst [707].

Table 28 [319]. 2-Ethylhexanol-1 through the hydroformylation of propylene with dicobalt octacarbonyl and catalyst additives
0.2 mole C_3H_6, 0.068 mole n-C_3H_7CHO and 0.4 millimole $Co_2(CO)_8$. Hydroformylation stage: 150 °C, 200 atm CO/H_2 = 1:2
Hydrogenation stage: 200 °C, 200–217 atm

| Catalyst | | | k | Product composition % | | | | | | | | |
Mg(OCH₃)₂	Pyridine	Fe(CO)₅	10^{-2}min⁻¹	C₃H₇CHO n	iso	C₄H₉OH n	iso	C₇H₁₃CHO	C₇H₁₅CHO	2-Ethyl-4-methyl-pentanol	2-Ethyl-hexa-nol	Residue
0.8	3.0	—	3.1	7.3	22.4	3.6	0.9	17.6	25.8	0	0	22.4
0.8	3.0	0.4	—	3.3	1.6	20.6	24.1	—	3.5	1.9	27.6	17.4
0.8	3.0	0.8	2.9	2.3	1.6	17.5	23.3	—	4.5	2.3	29.9	18.6
0.8	3.0	1.6	—	1.3	1.0	20.0	25.3	—	3.0	2.3	28.8	18.3
0.8	3.0	3.0	3.2	2.2	1.8	16.5	21.1	—	2.2	2.9	29.5	23.8
0.8	—	0.8	16.4	2.2	0.4	28.2	23.2	—	—	1.9	21.0	23.1
0.8	1.0	0.8	8.3	2.9	1.0	24.9	23.6	—	3.2	2.2	22.3	19.9
0.8	2.0	0.8	5.5	2.4	0.6	23.1	23.4	—	2.1	2.7	29.4	16.3
0.8	4.0	0.8	3.3	2.0	2.5	14.2	22.7	—	5.9	3.5	28.5	20.7
1.2	4.0	0.8	3.0	3.2	3.5	15.6	25.3	—	5.6	3.7	35.5	7.7
1.6	4.0	0.8	4.9	3.2	1.1	15.8	27.2	—	2.7	3.5	40.1	7.4

8. Ketone Formation under Hydroformylation Conditions

Hydroformylation is often accompanied by sequential or side reactions. Frequently found as sequential reactions are those already described: aldol condensation and acetal formation, Tishenko type reactions as well as oligomerization and polymerization. The hydrogenation of aldehydes to alcohols under hydroformylation conditions can also be included.

Hydrogenation of olefins to paraffins has already been mentioned as a side reaction of hydroformylation. The formation of formate esters is also frequently encountered, especially at higher temperatures (see the next section).

Ketone formation as a further major side reaction was noticed by Roelen in the discovery of hydroformylation with ethylene. The name "Oxo Reaction" was then given to the reaction of olefins with synthesis gas. It was found later that other olefins were much less susceptible to ketone formation and special conditions had to be used to obtain useful yields of ketones.

Cobalt in combination with Na_2O and Al_2O_3 [988] and Ru(III)acetylacetonate [989] were claimed to be superior catalysts.

Of special influence was the total pressure (table 29) and the ratio of olefin to carbon monoxide and hydrogen (table 30).

Table 29 [270]. *Influence of pressure and the ratio of H_2 to CO/C_2H_4 on the hydroformylation of ethylene*
140–150 °C

$H_2:CO:C_2H_4$	Press. (atm)	Reaction time (min)	Conversion (%)	% C_2H_5CHO in liquid product	% $C_2H_5COC_2H_5$ in liquid product
1 : 1 : 1	125	6.8	70.0	50.0	22.0
4 : 1 : 1	500	0.5	33.5	95.5	—
1 : 1 : 1	500	0.5	34.0	92.0	—
2 : 1 : 1	700	0.6	63.0	91.0	—

If the total pressure is low or the olefin excess high, the concentration of cobalt hydrocarbonyl and hydrogen will be reduced and, according to a mechanism proposed by Bertrand *et al.* [320], the acyl compound can react with the alkyl cobalt carbonyl instead of with hydrogen or cobalt hydrocarbonyl. The ketone is then formed as a result of this reaction.

From this mechanism, it follows that higher molecular weight olefins would react less, due to steric hindrance, which is in accord with the experimental results.

Table 30 [321]. *Variation of diethylketone yield with $C_2H_4/CO/H_2$ ratio*
Cobalt catalyst on carrier, 65.5 °C, 21 atm, fixed bed reactor

Mole ratio			g Liquid yield/m³	% $C_2H_5COC_2H_5$ in liquid product	% C_2H_5CHO in liquid product	% Conversion[a] to ketone
C_2H_4	CO	H_2				
1.0	1.0	1.1	430	24	16	16
1.5	1.0	1.0	555	53	5	35
2.4	1.0	1.1	635	78	3	57
3.8	1.0	1.0	460	80	2	57
5.7	1.0	0.9	267	82	2	48
1.0	1.0	1.9	495	20	22	19
1.8	1.0	1.8	646	42	11	33
2.9	1.0	2.1	524	62	9	50

[a] Calculated on basis of 2 C_2H_4/1 CO/1 H_2

$$\frac{\text{wt of recovered ketone x 100}}{\text{wt of charged mixture 2/1/1}}$$

$$H_2C=CH_2 + HCo(CO)_4 \longrightarrow H_3C-CH_2-Co(CO)_4 * \longrightarrow$$

$$H_3C-CH_2-\overset{\|}{\underset{O}{C}}-Co(CO)_3 \xrightarrow{CO} H_3C-CH_2-\overset{\|}{\underset{O}{C}}-Co(CO)_4$$

$$H_3C-CH_2-\overset{\|}{\underset{O}{C}}-Co(CO)_{3,4} + H_3C-CH_2-Co(CO)_4 \longrightarrow$$

$$H_3C-CH_2-\overset{\|}{\underset{O}{C}}-CH_2-CH_3 + Co_2(CO)_{7,8}$$

Alcohols can also function as hydrogen donors instead of molecular hydrogen; hence, the alcohol concentration must also be kept low to obtain good yields (table 31).

Table 31 [322]. *Formation of ketones by the hydroformylation of cyclohexene with isopropanol as hydrogen donor*

Alcohol/ cyclohexene	Yield (mole %)[a]		
	Dicyclohexyl-ketone	Cyclohexyl-carbinol	Ester
0.5	12	4	15
1.3	6	14	30
10.0	Traces	40	31

[a] Calculated on charged cyclohexene

* In these equations, the first three steps of the probable mechanism (see p. 5) have been condensed into the first step. It is not likely that the olefin adds pirectly to hydrotetracarbonyl.

Also in the reaction with alcohols, the yield with higher molecular weight olefins is lower (table 32).

Table 32 [322]. *Yield of dialkylketones with different olefins*

Olefin	Dialkyl ketone (mole %)	Aldehyde (mole %)	Ester (mole %)
Ethylene	57	5	21.5
Propylene	26	3	33.2
Isobutylene	Traces	3	56.0

Table 33 gives a summary of the preparation of ketones described above. The formation of cyclic ketones by the hydroformylation of dienes will be discussed in the chapter on ring closure reactions.

Table 33. *Ketone formation from olefins and synthesis gas*

Starting material	Reaction products	Yield (%)	Reference
Ethylene	Diethyl ketone	59.0	[270]
Propylene	Diisopropyl ketone, Isobutyraldehyde, Isopropyl alcohol		[270]
Butene-1 + HCo(CO)$_4$ in pentane	3,5-Dimethyl-4-heptanone, 3-Methyl-4-octanone, 5-Nonanone, n-valeraldehyde, 2-Methylbutanol (ketone: aldehyde = 9 : 1)		[320]
Ethylene + butanol-2	Diethyl ketone	64.5	[322]
Ethylene + methanol	Diethyl ketone	57.0	[322]
Cyclohexene + isopropanol	Bicyclohexyl ketone, Cyclohexyl carbinol, Isopropyl cyclohexylformate	12.0 4.0 15.0	[322]

9. Homogeneous Hydrogenation of Aldehydes under Hydroformylation Conditions

In the previous sections, it has been pointed out several times that the aldehyde group formed in the hydroformylation reaction can be subsequently hydrogenated to the hydroxyl group in the oxo reactor. It has been found that in most cases, if the stability of the cobalt hydrocarbonyl permits, that there is a homogeneous hydrogenation [323, 324], catalyzed by the hydroformylation catalyst. This can be illustrated [325] by the reactions (1–3). The equations (4–6) give the probable routes of the side reactions.

$$HCo(CO)_4 \rightleftharpoons HCo(CO)_3 + CO \tag{1}$$

$$RCHO + HCo(CO)_3 \rightleftharpoons R-\overset{\overset{\displaystyle H}{|}}{\underset{\downarrow}{C}}{=}O \rightleftharpoons R-CH_2-O-Co(CO)_3 \tag{2}$$
$$HCo(CO)_3$$

$$R-CH_2-O-Co(CO)_3 + H_2 \longrightarrow R-CH_2-O-CoH_2(CO)_3 \longrightarrow$$
$$R-CH_2-OH + HCo(CO)_3 \tag{3}$$

$$R-CH_2-O-Co(CO)_3 + CO \rightleftharpoons R-CH_2-O-Co(CO)_4 \tag{4}$$

$$R-CH_2-O-Co(CO)_4 \rightleftharpoons R-CH_2-O-\underset{\underset{\displaystyle O}{\|}}{C}-Co(CO)_3 \tag{5}$$

$$R-CH_2-O-\underset{\underset{\displaystyle O}{\|}}{C}-Co(CO)_3 + H_2 \rightleftharpoons R-CH_2-O-\underset{\underset{\displaystyle O}{\|}}{C}-CoH_2(CO)_3 \longrightarrow$$
$$R-CH_2-O-\underset{\underset{\displaystyle O}{\|}}{C}H + HCo(CO)_3 \tag{6}$$

The hydrogenation reaction requires a certain minimum temperature. This is generally above about 140–150 °C; below this temperature, only aldehyde is usually isolated after the hydroformylation.

The hydrogenation is first order in aldehyde, cobalt and hydrogen concentration. The carbon monoxide pressure has a similar effect to that observed in the hydroformylation reaction [325] (see fig. 10).

The reaction rate increases with increasing carbon monoxide pressure (until all the cobalt remains as carbonyl) and then falls off markedly. The decrease in rate of hydrogenation is substantially greater than that found for hydroformylation and is proportional to the second power of the CO partial pressure [325].

$$d(ROH)/dt = k\,(R'CHO)(Co)(H_2)(CO)^{-2}$$

According to this scheme, hydrogenation is effected by the addition of hydrogen to coordinatively unsaturated compounds with electron deficiency in the d-shell of the metal atom.

Equation (6) also contains a similar mechanistic influence and explains the frequently encountered side reaction of formate ester formation.

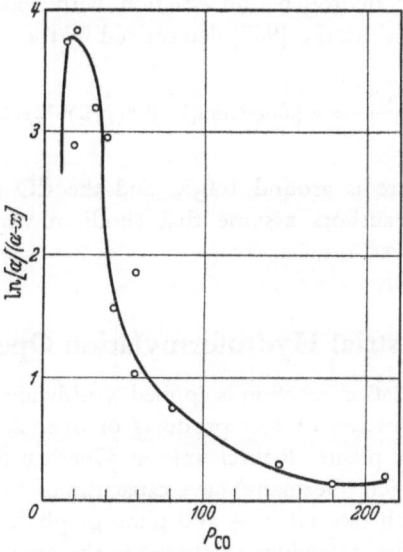

Fig. 10. Influence of the CO partial pressure (atm) on the reaction rate of hydrogenation of propionaldehyde in contact with dicobalt octacarbonyl
(0.2 mole % Co), ($P_{H_2} = 95$ atm), α = Initial concentration of aldehyde,
x = Concentration of formed alcohol.

The negative influence of the additional carbon monoxide pressure can
be explained with the assumption that the coordinatively saturated compound resulting from equation (4) does not react further. As equation (1)
indicates, the formation of the active cobalt hydrocarbonyl is also reduced
by higher CO pressure.

By use of selected reaction conditions, olefins can be converted directly
to alcohols without the isolation of the intermediate aldehyde.

In this manner, for instance, the Shell hydroformylation process has
been carried to commercial operation (see p. 22). The Shell catalyst (trialkyl phosphine/cobalt hydrocarbonyl), because of its high stability, is
especially useful for working at lower CO partial pressures. Rhodium
catalysts are also suitable for one-step alcohol syntheses; they also catalyze
the homogeneous hydrogenation of aldehyde groups [174, 253, 326, 327].
Rhodium catalysts also allow much higher reaction rates than cobalt
catalysts. The reaction goes especially smoothly when the rhodium is added
as carbonyl or in the form of its oxide.

According to Heil and Marko [328], aldehyde hydrogenation begins
to occur at CO partial pressures over 100 atm with rhodium chloride.
Rhodium chloride normally requires much more severe conditions to form
an active hydroformylation catalyst than rhodium oxide or rhodium carbonyl does [329].

The rate of the hydrogenation reaction with rhodium carbonyls is according to Heil and Marko [967] determined by the equation,

$$\frac{d\,(\text{alcohol})}{d\,t} = k\,[\text{aldehyde}][\text{Rh}]^{1/6}\,(p\text{H}_2)^{0.5}\,(p\text{CO})^{-0.3}$$

provided temperature is around 160 °C and the CO partial pressure at least 80 atm. The authors assume that rhodiumhydrocarbonyl acts as hydrogenation catalyst.

10. Industrial Hydroformylation Operations

The hydroformylation reaction is applied worldwide in large plants at a present world production of oxo products of over 2.7 million tons per annum. The largest plants, Ruhrchemie at Oberhausen (Germany) and BASF at Ludwigshafen (Germany) have capacities of well over 600 million pounds per year each (see table 34 and photograph facing page 72).

The operating plants produce aldehydes in the range C_3–C_{18} which are either hydrogenated as such to give the corresponding alcohols or subjected to aldol condensation prior to the hydrogenation. In the latter case the resulting alcohols contain double the number of carbon atoms as the aldehyde used for the aldol condensation (e. g. 2-ethylhexanol is made from two moles of n-butyraldehyde via 2-ethylhexenal).

Certain quantities of aldehydes are often oxidized for carboxylic acids in adjacent plants.

With only one exception, all industrially applied oxo processes more or less follow the technique which was developed by Ruhrchemie AG in Oberhausen/Germany [895–898, 1015] in co-operation with BASF. The exception is the process developed by Shell, which will be discussed later. The first industrial application dates back to 1940 when plant operations were started in the works of Ruhrchemie by a joint venture of Ruhrchemie, BASF and Henkel.

All existing oxo plants are operated fully continuously. In general they consist of the following sections:

1. Hydroformylation reactor
2. Catalyst removal section
3. Catalyst work-up and make-up section
4. Aldehyde distillation
5. Aldehyde hydrogenation reactor
6. Alcohol distillation

As mentioned earlier the plants may also contain aldolisation reaction and oxidation units.

As an example, see the flow sheet of the Ruhrchemie plant at Oberhausen, which is in principle the same for a number of plants constructed under license of Ruhrchemie (see table 34 on page 76/77). The flow sheets of the BASF process [340, 1037] and the Mitsubishi process [789] are very similar.

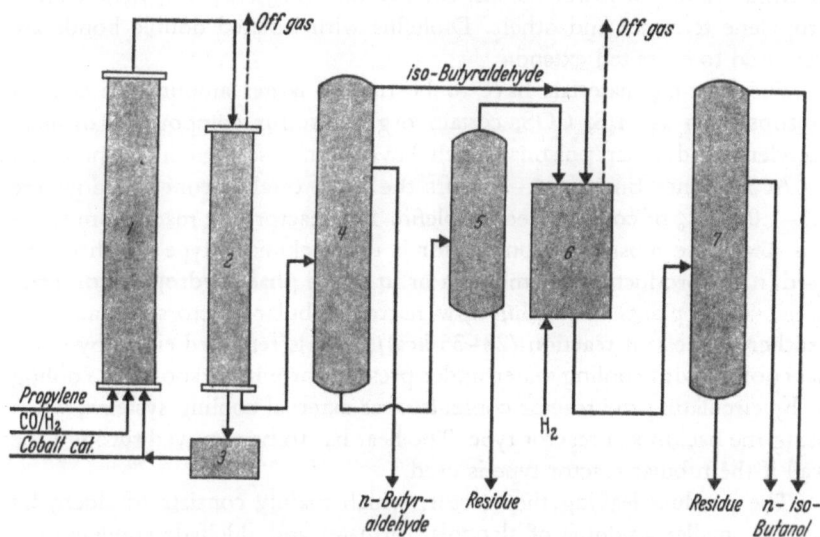

1 Hydroformylation reactor   5 Residue workup
2 Catalyst separator   6 Hydrogenation
3 Cobalt catalyst workup   7 Butanol distillation
4 Aldehyde distillation

Fig. 11 [792]. Flow sheet of Ruhrchemie oxo plant

The carbonyls or hydrocarbonyls of cobalt which are used as catalysts are generally formed in situ by feeding cobalt to the reactor in the form of metal oxide, hydroxide or salt of an organic or inorganic acid, either in solution or suspension in olefin, high boiling distillation residues or water. However, the carbonyls may also be formed in a small carbonyl generating reactor which is fed by the same catalyst precursors before they enter the hydroformylation reactor [161].

Usual reaction conditions for an oxo reactor are temperatures in the range of 110–180 °C and pressures of 200–350 atm with CO/H_2 ratios of 1:1 to 1:1.3.

Olefin and synthesis gas are fed to the reactor separately. However, in special cases such as ethylene hydroformylation they may be fed through one line. Solvents are generally not used in technical operations. Synthesis gas is either made by gasification of coal, by partial combustion of heavy oil or by steam reforming of methane or light hydrocarbons. Synthesis gas is desulfurized prior to its use in oxo operations. The following olefins are commonly used for industrial hydroformylations: ethylene, propylene, butylenes, crack olefins and olefins made by di- or trimerisation or co-di- or trimerisation of lower olefins, such as diisobutylene, propylene trimer, propylene tetramer and others. Diolefins with isolated double bonds are also used to a limited extend.

The starting materials have to be free of larger amounts of catalyst poisons such as H_2S, COS, certain organic sulfur compounds, oxygen, acetylene and other poisons which have been discussed in chapter 3.5.

At residence times of 1–2 hours the usual catalyst concentrations are 0.1–1.0 wt-% of cobalt based on olefin. The reactors are made from stainless steel. The most common reactor is the backmixed type which is also used in the production of ammonia or in liquid phase hydrogenation reactions. Some plants have plug-flow narrow tubular reactors instead. The exothermic heat of reaction (28–35 kcal/mole) is removed either by internal cooling with cooling water under pressure or with evaporative cooling or by circulating the reactor contents over external cooling systems, when using the backmixed reactor type. The heat has to be removed through the wall if the tubular reactor type is used.

The product leaving the reactor, which mainly consists of aldehydes besides smaller amounts of alcohols, formates and aldehyde condensation products, is fed to the decobalting section.

Two different decobalting methods are used in a number of variations. One is the thermal decomposition of the carbonyls after releasing the product from the reactor with simultaneous reduction of the CO-partial pressure. The decomposition is effected by recycling cobalt-free hot reaction product or by introducing hot water or steam [838]. An inert stripping gas may be applied simultaneously. The solid cobalt metal, -oxide, or -hydroxide separating is mechanically removed [749, 839].

The other is the removal of cobalt from the crude oxo product by treatment with chemicals. This may be achieved by reaction with aqueous acids or salts with or without simultaneous application of an oxidizing agent such as air. The resulting cobalt-containing acid solution is worked up and recycled to the reactor [840, 841]. Thus, Union Carbide removes cobalt at elevated temperatures with sulfuric or acetic acid [842]. A variation of this method was developed by Kuhlmann who extracts cobalt hydrocarbonyl from the crude oxo product by a dilute sodium bicarbonate solution. Subsequent acidification of the water layer reforms the hydrocarbonyl which is recycled to the oxo reactor as such [843].

The demetallization may be accelerated by oxidizing agents such as oxygen or air [780]. BASF, e.g. feeds the 120 °C hot oxo raw product to a

Part of the Ruhrchemie Oxo Plant at Oberhausen-Holten

small reactor in which it is mixed with dilute acetic acid and two oxidation equivalents of air per mole of cobalt contained in the crude oxo product. The decobaltizer is kept under a pressure of 10 atm [844].

The aldehyde may be sent to the hydrogenation reactor without distilling it in advance. However, if alcohols of extremely high purity are desired, an aldehyde distillation may be recommended.

For corrosion reasons the aldehydes are distilled in columns made from stainless steel.

The hydrogenation of oxo aldehydes is generally done in fixed bed reactors either in the liquid or in the gas phase, copper or nickel catalysts being applied.

Many oxo plants contain an additional aldolization section, e. g. for the manufacture of 2-ethylhexanol from n-butyraldehyde [893, 894].

As mentioned earlier the only oxo process which differs considerably from the Ruhrchemie technique is the method developed by Shell [892], see fig. 12.

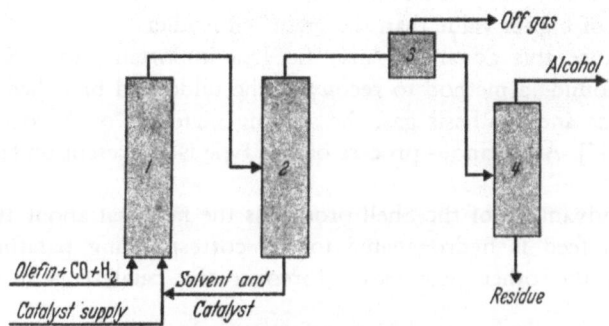

1 Hydroformylation and hydrogenation reactor
2 Pressure distillation
3 Pressure reduction and gas separator
4 Alcohol rectification

Fig. 12. Flow sheet of Shell oxo plant

Shell uses a special catalyst system (see chapter 3.6 on catalyst modifiers). Cobalt salts of organic acids (e. g. Co-2-ethylhexanoate), trialkyl phosphines (e. g. tributyl phosphine) and alkali, such as KOH, are fed to the oxo reactor as catalyst precursors. Under the conditions of the Shell process $HCo(CO)_3PR_3$ is formed from these compounds which is the catalyst active in the Shell method. Since this catalyst is thermally more stable than $HCo(CO)_4$ the oxo reactor needs only a pressure around 100 atm, versus 200–300 atm in the other processes. However, compared with $HCo(CO)_4$ the catalyst is less reactive. Thus at equal concentrations even at 180 °C it gives only $1/2$ to $1/5$ of the rate of reaction which can be reached with $HCo(CO)_4$ at 145 °C. This means that an approximately 5 to 6 fold increase in reactor volume is required for the same throughput as with $HCo(CO)_4$.

Since $HCo(CO)_3PR_3$ is not only a hydroformylation but also a good hydrogenation catalyst the aldehydes formed are hydrogenated to alcohols in the Shell reactor when working around 180 °C. The catalyst is so stable that the alcohols can be flashed from the catalyst and high boiling residue in a pressure distillation unit. The crude alcohols are purified in distillation units while the catalyst containing heavy ends of the pressure distillation is recycled to the oxo reactor [974]. The fact that the hydrogenation can be achieved in the oxo reactor makes up for the lower activity of the catalyst in the oxo step.

In many cases the Shell catalyst gives a higher n/iso ratio of the aldehydes formed from straight chain olefins (88 : 12 n-butanol/isobutanol from propylene versus 80 : 20 n-butyraldehyde/isobutyraldehyde in the Ruhrchemie process) which is an advantage since straight chain compounds are generally of higher value than the branched isomers.

In future this advantage may be less important since Ruhrchemie recently found a method to reconvert the undesired branched aldehydes into olefins and synthesis gas, the starting materials of the oxo synthesis [1015, 1046]. A continous process of this type is at present under development.

A disadvantage of the Shell process is the fact that about 10–15 % of the olefin feed is hydrogenated to the corresponding paraffins against 1–3 % in the other processes. Moreover the catalyst is more expensive.

On the basis of the data available so far it looks as if this process is well suited for the single-step production of alcohols but less suited if aldehydes are the desired reaction products, which is the case in the 2-ethylhexanol manufacture and similar aldol routes, in the manufacture of carboxylic acids and some other operations (as to the possibility of making a certain amount of aldol product along with the lower molecular alcohols in the Shell process, see chapter 7).

A second oxo method which differs from the classical route — the gasphase hydroformylation — is still in the research stage in Russia [900, 901].

Special methods are used by the different manufacturers for the work-up of the heavy ends resulting from side reactions in the oxo reactor and obtained in the aldehyde distillation column.

Some manufacturers hydrogenate this residue under severe conditions to make additional alcohol [845, 846, 1037]. Others work-up for aldehyde and some treat the bottoms with alkali, e.g. with NaOH at high temperature (260–350 °C) in order to make carboxylic acids [339, 847–851].

11. Application and Economics

Primary alcohols are by far the most important products made via the hydroformylation route [852]. The lower alcohols (C_3–C_5) are used as solvents [853, 854, 1032], the range from C_4–C_{13} mainly for the production of plasticizers, especially phthalates [853, 855] and the higher homologs above C_{13} for detergent manufacture [853]. Nearly three-fourths of the oxo output is used for plasticizer manufacture [877, 904].

As an interesting new use the application of butanol for the manufacture of drinking water from salt water may be mentioned [343]. As to other uses see the booklet of Hoechst-Ruhrchemie "Produkte aus der Oxo-Synthese — Aldehyde, Alkohole, Säuren —" [853].

Especially in the range of C_8–C_{13} the oxo route holds the dominating position since the other routes, like the aldol route starting from acetaldehyde, the hydrogenation of fatty acid esters or the Ziegler alcohol route, can hardly compete pricewise [342, 899].

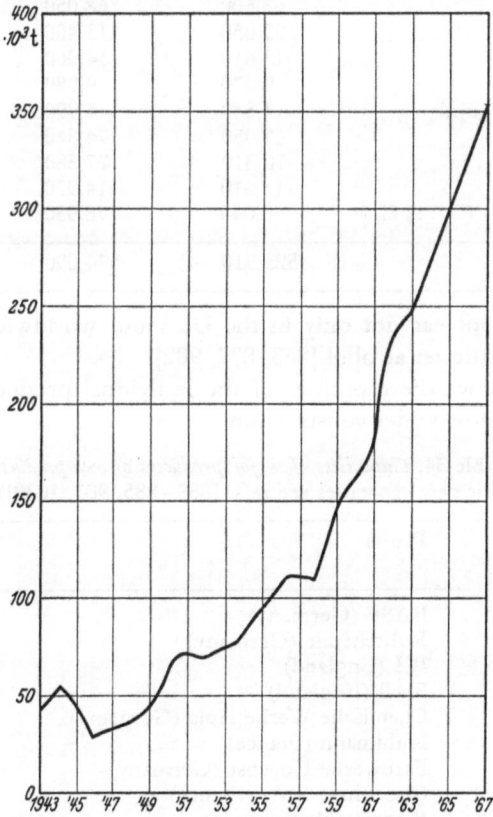

Fig. 13 [787, 1054]. Growth of dialkyl phthalate production in USA from 1943–1967

The increase of oxo alcohol production is closely related to the growth rate of PVC since soft PVC contains approx. 30–40 wt-% of plasticizers which are mainly made from oxo alcohols and phthalic anhydride. As to the rapid growth of dialkyl phthalate manufacture in the USA, see fig. 13.

In 1965 the western countries produced around 2.8 million tons of PVC containing approximately 800,000 tons of plasticizers [340]. A yearly growth rate of 10–15% is estimated both for PVC and for oxo products. From 1964 to 1966 a yearly growth rate of over 20% was reported [863, 877].

The production in the USA consists of the following products [1031].

	1968	1969	1973 (Estimated)
		In tons/annum	
Propionic acid	13,610	14,520	18,140
n-Propanol	6,800	7,710	9,070
n-Butanol	90,720	95,260	117,940
Isobutanol	63,500	68,050	99,790
2-Ethylhexanol	102,050	113,400	167,830
Butyric acid	13,610	14,060	18,140
C_5-Compounds	9,070	9,980	11,340
Isohexyl-compounds	4,540	4,990	6,800
Isooctanol	68,050	74,840	108,860
Isodecanol	70,310	77,380	113,400
Tridecanol	13,610	14,970	22,680
Other Products ($C_3 \ldots C_{14}$)	68,040	78,930	122,470
Total	523,910	574,090	816,460

2-Ethylhexanol has not only in the USA but worldwide become the number-one plasticizer alcohol [863, 877, 902].

Table 34 shows the capacities of the individual producers which are either producing or under construction.

Table 34. *Capacities of major producers of oxo products* (for new capacities see refs. [859–885, 903, 1030])

	Plants	Oxo products in 1000 tons
Europe	BASF (Germany)	370
	Ruhrchemie (Germany)	320
	ICI (England)	200
	Shell (England)	170 [a]
	Chemische Werke Hüls (Germany)	100 [e]
	Kuhlmann (France)	100
	Farbwerke Hoechst (Germany)	56 [d]
	Oxochimie S.A. (France)	56 [d]
	Konam (Netherlands)	50 [g]
	Österreichische Stickstoffwerke (Austria)	45 [e]

Table 34 (continued)

	Plants	Oxo products in 1000 tons
Europe	Celanese S. P. a. (Italy)	40
	Riminicu Vilcea (Rumania)	37 d
	Technoimport (Bulgaria)	30 c
	BASF-Arrahona (Spain)	28 c, e
	Montedison (Italy)	25
	Chemapol (Czechoslovakia)	20 f
		1,647 b
America	Union Carbide Corp. (USA)	180
	W. R. Grace/Corco (Puerto Rico)	110 e
	Dow Badische Co. (USA)	100 a, e
	Shell Chemical Co. (USA)	91
	Texas Eastman Co. (USA)	90
	Gulf Oil (Canada)	56 c, d
	Enjay Chemical Co. (USA)	55
	BASF (Canada)	45 c
	U. S. Steel Chemicals Inc. (USA)	32
	Gulf Oil Corp. (USA)	23
	Getty Oil/Houdry (USA)	18
		800
Asia and Australia	Kyowa Yuka (Japan)	71 d
	Chisso (Japan)	52 d
	Mitsubishi Chem. Ind. Ltd (Japan)	50
	Nissan Petrochem. Co. (Japan)	40 g
	C. S. R. Chemicals Pty Ltd. (Australia)	24 d
	Tonen Petrochem. Co. (Japan)	20
	Dai Nippon INK (Japan)	20 g
	Daikyowa Petrochem. Co. (Japan)	12
	Nocel (India)	8
		297
	Total	2,744

a Under construction
b In Russia 3 plants are said to be on stream; however, capacities are unknown
c Planned
d License of Ruhrchemie
e License of BASF
f License of Mitsubishi
g License of Kuhlmann

The outlook for further increase of the oxo capacities is bright, although many people foresee a shortage of propylene which is the key product among the starting materials [886–890, 1033, 1038–1040]. In Western Europe the oxo-producers have already become the biggest propylene consumers [890].

II

Metal Carbonyl Catalyzed Carbonylation
Reppe Reactions

1. General Remarks

Carbon monoxide reacts with unsaturated compounds (or components which are able to form unsaturated compounds) and a nucleophilic compound containing a mobile H-atom in the presence of certain metal carbonyls to yield carboxylic acid derivatives. This reaction is called carbonylation or, if acetylenes or olefins are reacted with carbon monoxide and water, sometimes is named hydrocarboxylation.

Carbonylation may also be achieved by insertion of carbon monoxide into an existing bond, as for instance in the formation of anhydrides from carboxylic acid esters, of esters from ethers, or acids from alcohols.

The metal carbonyl catalyzed reaction was discovered by W. Reppe and followed up from 1938–1945 by Reppe and his co-workers in the laboratories of BASF. The results laid down in patent applications remained unpublished until the end of World War II and a detailed report by the inventors was published only many years later [1, 239, 345–349]. Some other reviews appeared later [350–354].

$$HC{\equiv}CH + CO + HOR \longrightarrow H_2C{=}CH{-}\underset{\underset{O}{\|}}{C}{-}OR$$

$$HC{\equiv}CH + CO + HNR_2 \longrightarrow H_2C{=}CH{-}\underset{\underset{O}{\|}}{C}{-}NR_2$$

$$H_2C{=}CH_2 + CO + H_2O \longrightarrow H_3C{-}CH_2{-}COOH$$

$$H_3COH + CO\ (+ H_2O) \longrightarrow H_3C{-}COOH$$

$$H_2C{=}CH{-}CH_2{-}Cl + CO + ROH \longrightarrow \begin{array}{l} H_2C{=}CH{-}CH_2{-}COOR \\ H_3C{-}CH{=}CH{-}COOR + HCl \end{array}$$

There is a wide variety of reactants in this reaction. Suitable unsaturated starting materials or unsaturated compound-forming components are: acetylenes, olefins, alcohols, cyclic and non cyclic ethers, epoxides, acetals, esters, saturated aldehydes, lactones and halides. Water, alcohols, ammonia, amines, mercaptans and carboxylic acids, e. g. are used as nucleophilic compounds with mobile hydrogen atoms.

The reactions proceed either under pressure with gaseous carbon monoxide in the presence of catalytic amounts of metal carbonyls, carbonyl forming metals or their derivatives, or with stoichiometric amounts of metal carbonyls as CO-donors at atmospheric pressure. Addition of acids or acid forming halogens accelerates the reaction. In case of the stoichiometric, pressureless method, the addition of acids is essential.

The stoichiometric method, in which $Ni(CO)_4$ is preferably used as CO-source, has the advantage that no pressure vessels are required and low reaction temperatures (appr. 40 °C) may be applied. A drawback is that large amounts of solid nickel salts have to be separated without losses and have to be worked up for $Ni(CO)_4$.

Since a nickel work-up without any losses is hard to realize [355] in most cases, the catalytic process is more economic than the stoichiometric procedure (for recovery of $Ni(CO)_4$, see W. Reppe and W. Schlenk, German Pat. 753618 [779].

2. Reaction Mechanism

The reaction mechanism remained unclarified for nearly 25 years and some details are not certain to-day. The mechanisms proposed for the early period of the reaction, like the intermediate formation of ketenes [3, 6] or cyclopropanones [345], were shown to be incorrect [3, 6, 356]. The hypothesis of M. Almasi et al. [357], who assumed that first a carbonium ion is formed from the unsaturated component and a proton which then reacts with the metal carbonyl, is not very likely since acetic acid, e. g., which can successfully be applied in carbonylation, is not able to protonize an alkyne under conditions used in the carbonylation [358–360].

In the case of cobalt catalysts the reaction very likely proceeds according to equations (1–3), as proposed by Heck [361]. This mechanism is closely related to that of the hydroformylation reaction (see page 5).

Equations (1) and (2) may be broken down into a number of consecutive steps (see page 5).

$$H_2C{=}CH_2 + HCo(CO)_4 \longrightarrow H_3C{-}CH_2{-}Co(CO)_4 \tag{1}$$

$$H_3C{-}CH_2{-}Co(CO)_4 \xrightarrow{\text{CO}} H_3C{-}CH_2{-}\underset{\underset{O}{\|}}{C}{-}Co(CO)_4 \tag{2}$$

$$H_3C{-}CH_2{-}\underset{\underset{O}{\|}}{C}{-}Co(CO)_4 \xrightarrow{\text{ROH}} H_3C{-}CH_2{-}COOR + HCo(CO)_4 \tag{3}$$

Where $Ni(CO)_4$ is used, Heck proposed a mechanism following equations (4–8a), formulated with the example of olefins:

$$HX + Ni(CO)_4 \rightleftharpoons HNi(CO)_2X + 2\,CO \qquad (4)$$

$$HNi(CO)_2X + RCH=CH_2 \longrightarrow RCH_2-CH_2-Ni(CO)_2X \qquad (5)$$

$$\searrow \quad \begin{array}{c} R-CH-CH_3 \\ | \\ Ni(CO)_2X \end{array}$$

$$RCH_2-CH_2-Ni(CO)_2X + CO \longrightarrow RCH_2-CH_2-\underset{\underset{O}{\|}}{C}-Ni(CO)_2X \qquad (6)$$

$$RCH_2-CH_2-\underset{\underset{O}{\|}}{C}-Ni(CO)_2X \xrightarrow{2\,CO} RCH_2-CH_2-CH_2-COX + Ni(CO)_4 \qquad (7)$$

$$RCH_2-CH_2-COX + R'OH \longrightarrow RCH_2-CH_2-\underset{\underset{O}{\|}}{C}-OR' + HX \qquad (8)$$

or

$$RCH_2-CH_2-\underset{\underset{O}{\|}}{C}-Ni(CO)_2X \xrightarrow{R'OH} RCH_2-CH_2-COOR' + HNi(CO)_2X \qquad (8a)$$

However, it must be mentioned that the existence of nickel hydrocarbonyls is not yet confirmed since there are only indirect hints as to their existence [362–364]. The only compound in this series which could definitely be established is $(HNi(CO)_3)_2 \cdot 4NH_3$.

With acetylenes the reaction may proceed through (4a) and (5a), as proposed by R. W. Rosenthal et al. [167].

$$HC\equiv CH + Ni(CO)_4 \longrightarrow \begin{array}{c} H \\ C \\ \| \quad\diagdown \\ \quad\quad Ni(CO)_2 + 2\,CO \\ C \diagup \\ H \end{array} \qquad (4a)$$

$$\begin{array}{c} H \\ C \\ \| \quad\diagdown \\ \quad\quad Ni(CO)_2 + HX \\ C \diagup \\ H \end{array} \longrightarrow H_2C=CH-Ni(CO)_2X \qquad (5a)$$

(4a) very likely follows an Sn_1 mechanism with dissociation of $Ni(CO)_4$ into $Ni(CO)_3$, followed by complexing of $Ni(CO)_3$ with the acetylenic compound. This assumption was also made by Ehrreich et al. [365] in their reaction proposal for the mechanism of carbonylation. $Ni(CO)_3$ was first described by Thomson [366]. A reaction mechanism with an intermediate formation of $Ni(CO)_3$ would also well explain the hindrance of the reaction by high CO-partial pressure (see also page 86). The induction period frequently observed in the carbonylation using nickel carbonyl and its length, which depends on reaction temperature, pressure, solvent and structure of the alkyne [367], as well as the acceleration of the reaction by irradiation with UV-light [358], also gives rise to the possibility that a species with less than four CO-ligands is the active catalyst rather than $Ni(CO)_4$.

Table 35 [146]. *Induction period and half-time values of isomeric n-octynes in the Ni-carbonyl-catalyzed hydrocarboxylation under various reaction conditions*

n-Octyne	Temp. (°C)	Initial pressure (atm)	$t_1/_2$ (min)	I (min)
−(1)	200	120	2.5	−
−(1)	200	200	6	−
−(1)	200	258	46	−
−(3)	200	120	13	20
−(3)	200	200	48	115
−(2)	200	200	48	70
−(4)	200	200	48	20
−(2)	200	258	98	120

Jones *et al.* [369] showed that small amounts of pyridine, which catalyze the $Ni(CO)_4$ dissociation, shorten the induction period. Ehrreich *et al.* [365] demonstrated that the induction period was shortened to $1/_{10}$ of the usual time if a stream of nitrogen was passed through the reaction vessel after the $Ni(CO)_4$ had been introduced. Removal of the CO-gas phase also favors the formation of $Ni(CO)_3$.

Addition of compounds containing divalent sulfur, selenium or tellurium also shortens the induction period and increases the rate of reaction. These compounds are highly active if nickel or cobalt halides are used as catalyst precursors. The increase in the rate of reaction may well be explained by the faster regeneration of the active intermediate nickel carbonyl complexes [370].

The dependence of the induction period and half-time values on the pressure is well demonstrated by table 35 and fig. 14, showing the catalytic hydrocarboxylation of n-octynes in the presence of $Ni(CO)_4$.

As an alternative to the discussed mechanism, Heck proposed the formation of a halide from the unsaturated compound and the acid present in the reaction mixture. This halide is then thought to react with $Ni(CO)_4$ to form an alkyl nickel dicarbonyl halide according to equation (4b).

$$RX + Ni(CO)_4 \longrightarrow RNi(CO)_2X + 2\,CO \qquad (4b)$$

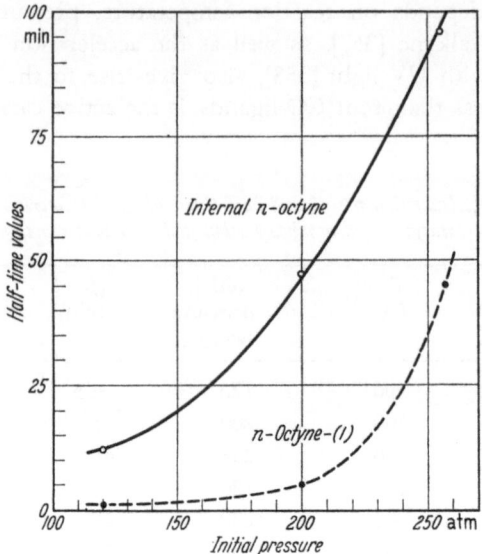

Fig. 14 [146]. Half-time values of the hydrocarboxylation of isomeric n-octynes in dependence on the initial pressure at 200 °C in presence of catalytic amounts of Ni-carbonyl

A mechanism which is slightly modified but in principle in line with the proposal of Heck was proposed by v. Kutepow, Himmele and Hohenschutz [371, 372] for the reaction of methanol with CO/H_2 [373], in the presence of a cobalt iodide catalyst: (9)–(14).

$$2\,CoI_2 + 2\,H_2O + 10\,CO \longrightarrow Co_2(CO)_8 + 4\,HI + 2\,CO_2 \quad (9)$$

$$Co_2(CO)_8 + H_2O + CO \longrightarrow 2\,HCo(CO)_4 + CO_2 \quad (10)$$

$$CH_3OH + HI \longrightarrow CH_3I + H_2O \quad (11)$$

$$CH_3I + HCo(CO)_4 \rightleftharpoons CH_3Co(CO)_4 + HI \quad (12)$$

$$CH_3Co(CO)_4 \longrightarrow CH_3COCo(CO)_3 \underset{-CO}{\overset{+CO}{\rightleftharpoons}} CH_3COCo(CO)_4 \quad (13)$$

$$CH_3COCo(CO)_{3,4} + H_2O \longrightarrow CH_3COOH + HCo(CO)_{3,4} \quad (14)$$

The mechanism discussed here may help to understand the carbonylation reaction. Its proof or disproof must be left to future experiments.

As in the hydroformylation reaction, isomeric mixtures of the reaction products are often obtained. However, since the different starting materials show a different behaviour in this respect, these isomerizations will be discussed in the special chapters dealing with the single classes of starting compounds.

3. Catalysts

Active catalysts are nickel, cobalt, iron, rhodium, ruthenium and palladium, as well as their salts, carbonyls or hydrocarbonyls.

These elements are used as carbonyls or in form of metals, salts, complex salts or oxides depending on the desired reaction.

As mentioned in the chapter on the reaction mechanism, the anion, especially of Ni-salts, is important in affecting the reaction course. The catalytic efficiency of the nickel halides strongly increases in the series: fluoride, chloride, bromide, iodide [374–376]. The molar ratio of cobalt or nickel to iodine is also very important [414]. As in the hydroformylation reaction, metal carbonyls substituted by phosphine ligands are very reactive [377, 1009], and especially modified rhodium and palladium catalysts [1021, 1045] allow reactions under mild conditions. Thus, the nickel bromide triphenylphosphine allyl bromide complex shows an increased reactivity in the carbonylation of acetylenes. On the other hand, carbonyls substituted by phosphine ligands are also readily soluble in the reaction mixture [345, 377].

Selenium [379], copper [146, 377, 380, 381], silver [380] and zinc [380] are recommended as promoters in reactions catalyzed by $Ni(CO)_4$. Selenium is said to increase the yield and the lifetime of the catalyst and to prevent the formation of polymeric by-products.

A general description of the activity of the various metal carbonyls cannot be given, as it depends decisively on the compound which is to be carbonylated. Whereas the carbonylation of acetylenes and olefins is most effectively catalyzed by nickel, cobalt is preferred for the carbonylation of alcohols, ethers and epoxides. Often various carbonyls result in different reaction products with the same starting material. Thus, in the catalytic procedure, acetylene and carbon monoxide react in the presence of water with $NiBr_2$/$CuBr_2$-catalysts to give acrylic acid in high yield, using tetrahydrofuran or acetone as solvent [382]. Under the same reaction conditions with the same catalyst, diacrylic or triacrylic acid, instead of acrylic acid, is obtained as the main product when water is used as solvent [383].

$$H_2C=CH-COO-CH_2-CH_2-COOH \text{ (diacrylic acid)}$$

$$H_2C=CH-COO-CH_2-CH_2-COO-CH_2-CH_2-COOH \text{ (triacrylic acid)}$$

These acids are formed sequentially from acrylic acid and water via β-hydroxypropionic acid as intermediate. Yields are dependent on the amount of water present in the reaction mixture.

If the same reaction is carried out at temperatures between 250–270 °C and slightly increased CO-partial pressure, succinic acid is obtained as the main product [384]. High acetylene partial pressure in the acrylic ester synthesis from acetylene, carbon monoxide and alcohol in the presence of triphenylphosphine nickel halogenide catalysts results in formation of heptatriene-2.4.6-acid ester-1 as main product, by subsequent reaction of the acrylic acid ester formed in the first stage of the reaction [385]. Therefore, acrylic acid ester, which is thought to be the intermediate, may advantageously be used as solvent.

If the reaction of acetylene, carbon monoxide and alcohol in the presence of $Ni(CO)_4$ is carried out in strong acid medium with only a low CO-partial pressure, the main reaction products are unsaturated dicarboxylic acid esters of the type $ROOC-(CH=CH)_n-COOR$, n being mainly 3 or 4 [386, 387].

The acrylic acid ester initially formed in the reaction of acetylene, carbon monoxide and water/alcohol mixtures reacts to give succinic acid diester [205, 226, 388–400] in a secondary reaction if cobalt catalysts are used.

A mixture of cis-trans-4,5-dihydroxyoctatriene-2,4,6-diacid-1,4;8,5-dilactone is obtained in good yield with dioxane or cyclohexanone as solvent in the presence of cobalt catalyst at high CO partial pressures (e.g. 600 atm) [401].

Under the same reaction conditions with slightly increased temperatures (110–170 °C) and iron carbonyl as catalyst, p-quinone was found to be the main reaction product [402]. If the same reaction is carried out with a slightly increased amount of water hydroquinone is obtained in good yields [399, 400, 402]. As with the conversion of acetylene with carbon monoxide and a third component, in the reaction of ethylene, carbon monoxide and water or alcohol, different reaction products may also be obtained by altering the reaction conditions and catalysts.

Ethylene reacts with carbon monoxide and water in the presence of nickel carbonyl to give propionic acid in high yield. If care is taken to maintain a high concentration of propionic acid in the reaction mixture and the temperature, which is normally 300 °C in the propionic acid synthesis, is decreased to 240 °C propionic acid anhydride is formed in high yield in the presence of $Ni(CO)_4$. Propionic acid ethyl ester is the main product in the reaction of ethylene, carbon monoxide and water (low water concentration must be applied) with cobalt carbonyls instead of $Ni(CO)_4$. The conversion of ethylene with carbon monoxide in dilute alkaline medium with the aid of potassium nickel cyanide gives propionyl propionic acid [403–405]. At higher temperatures and without pH correction in the same reaction mainly polyketones with the sequences $-(CH_2-CH_2-CO)-$ are formed. If the reaction is carried out in absence of water or alcohols and in presence of palladium iodide as catalyst, a mixture of hexenolide isomers is the main product. Colorless polyketones of the same structure are obtained if an excess of ethylene is treated with carbon monoxide in the presence of complex palladium salts as catalysts in an alcoholic hydrogen halide solution at 100 °C and 700 atm [406].

Ethylene and carbon monoxide react to give n-propanol as the main product in the presence of iron carbonyl and organic bases such as N-n-propyl pyrrolidone [406].

This review shows that, by variation of operation conditions, alteration of catalysts and additional components as well as by variation of solvents and of concentrations, very different reaction products may be obtained.

The selection of the catalyst system will often be determined by the process technique used. In the catalytic processing of acrylic acid-n-butyl ester synthesis from acetylene, carbon monoxide and n-butanol with nickel halogenide, troublefree continuous operation could not be achieved.

Better results were obtained with catalysts of the type of quaternary phosphonium nickel halogenides or the corresponding ammonia compounds [408, 409]. Nickel bromide and nickel iodide are only soluble in butyl acrylate/butanol mixtures in amounts which are insufficient for the maintenance of the catalytic process. As mentioned in patents [408, 409], tertiary and quaternary complex compounds supply the required concentration of nickel and halogen ions for a continuous reaction. Many of the carbonylations, especially the stoichiometric reactions with $Ni(CO)_4$ are carried out in the presence of water and acids. Jones [369] investigated the efficiency of the monobasic acids and showed hydrochloric acid and acetic acid to be of the same efficiency, whereas trichloro acetic acid is inefficient. He concludes that not the protons of the acids but the anions are of importance in the reaction mechanism. Apart from hydrochloric acid and acetic acid [345], sulfuric acid, aqueous phosphoric acid, formic acid [367] and monochloroacetic acid are suited. A number of other organic acids [367] are ineffective. Acetic acid may be used in less than the equivalent amount because the unsaturated acid formed in the hydrocarboxylation [367] may replace acetic acid.

Oxidizing agents such as oxygen and nitrobenzene are catalyst poisons [365]. Surprisingly, CCl_4 is also a catalyst poison in the carbonylation of acetylene with $Ni(CO)_4$ [365].

The best catalyst systems for each starting material will be discussed in the following chapters on the individual starting materials.

4. Influence of Pressure and Temperature

The necessary temperatures and pressures of the carbonylation reactions are largely determined by the unsaturated starting material and the catalysts applied. In the pressureless stoichiometric carbonylation of acetylenes with $Ni(CO)_4$, temperatures of 35 to 80 °C are sufficient. In the catalytic procedure with $Ni(CO)_4$ temperatures of 120 to 220 °C and pressures around 30 atm are normally applied. In special cases lower temperatures may be used, e.g. in the acrylic acid and acrylic acid ester syntheses, which proceed in the presence of acetylacetone or triphenyl phosphine in the catalytic procedure at temperatures of 20 to 30 °C lower than usual in the catalytic method, with the same reaction velocity at the same concentration of nickel bromide.

Most of the catalytic reactions require a certain minimum temperature. Irrespective of pressure, no reaction occurs below this temperature. This temperature, e. g., is 180 to 183 °C for n-octyne and 198 to 200 °C in the case of internal octynes [146]. The effect of temperature on the reaction velocity at different catalyst concentrations was investigated by a number of authors [410–413].

However, the reaction rate is determined less by the reaction temperature than by the mass transfer of carbon monoxide and acetylene in the reactor [371].

The stoichiometric carbonylation of olefins and their derivatives may also be achieved without pressure at room temperature (in special cases even at lower temperatures). With catalytic amounts of cobalt hydrocarbonyl, temperatures between 100 and 260 °C and pressures from 30 upward to 900 atm are applied. In the carbonylation of saturated alcohols, cobalt catalysts are preferred, which are already highly active at 180 °C, whereas nickel carbonyl requires a reaction temperature of 280 °C to give the same results [371].

In the cobalt catalyzed carbonylation, as well as in hydroformylation, an inhibition is observed with increase of the carbon monoxide pressure [365, 415–417]. An even larger inhibition effect is noted in nickel catalyzed reactions [411]. An explanation of these facts is given in the chapter on the reaction mechanism.

Isomer mixtures are often formed in the carbonylation reaction. The formation of straight chain alkyl carbonyl compounds is favored by high CO partial pressure and low reaction temperatures, while low CO pressure and high reaction temperature favor the formation of branched alkyl carbonyl compounds. If the charge distribution in an olefin is changed, e. g. by substituents (as in α,β-unsaturated esters), the effects may be reversed. In these cases high CO-partial pressure and low temperature favor formation of branched reaction products whereas straight chain products are obtained if low partial pressure and high temperatures are applied. The isomer distribution will be discussed in detail in the individual chapters [146, 365, 371, 415–417].

5. Solvents for the Carbonylation Reaction

The stoichiometric syntheses with $Ni(CO)_4$ are carried out in the presence of aqueous acids such as hydrochloric acid, phosphoric acid, acetic acid or acid forming nickel salts. Primary, secondary and tertiary alcohols, ketones, ethers, esters or pyridines are used as solvents [365]. However, with the exception of pyridine, the solvents have no remarkable influence on the reaction course. Compared with ethanol, tertiary butanol, acetone and anisol extend the inhibition time. With pyridine the inhibition time is shortened considerably (for reasons see the section on reaction mechanism).

The presence of at least stoichiometric amounts of water is important, otherwise the yield decreases considerably with simultaneous hydrogenation of large amounts of the starting material [365]. Larger amounts of water are favorable in the catalytic syntheses of carboxylic acids from saturated alcohols [373, 411–413].

On the one hand water suppresses the conversion of starting material and reaction products to esters, and on the other hand the conversion of carbon monoxide to carbon dioxide by the water gas reaction [371] is inhibited.

$$CO + H_2O \longrightarrow CO_2 + H_2$$

The cobalt carbonyl catalyzed carbonylation of unsaturated compounds may be carried out in the absence of water, with the exception of the production of saturated carboxylic acids. In the anhydrous procedure nonpolar solvents such as paraffins or aromatic hydrocarbons can be used. However, preferable solvents are alcohols, ketones, ethers, esters or nitriles.

In the syntheses of low molecular weight carboxylic acids, water has a positive effect on the reaction course while in the production of high molecular weight carboxylic acids starting from the corresponding olefins, the presence of water is a disadvantage as the catalyst components will preferentially be dissolved in the aqueous layer. However, an improvement can be achieved by addition of solvents such as carboxylic acids.

The carbonylation reaction is often successfully carried out in a homogeneous phase in alcohols as solvents.

Favorable reaction conditions in the carbonylation of molecules with olefinic double bonds are obtained if the starting compound contains a hydrophilic group, as e. g. in oleic acid, where the solubility of water in the organic layer is increased by the presence of the carboxyl group. Thus, a troublefree reaction course is achieved even in continuous processing. In this case even dicarboxylic acids, instead of the diesters, are obtained in high yield.

6. Carbonylation of Various Structures

6.1. Acetylenes and Their Derivatives in the Presence of Water

The catalytic carbonylation of acetylenes or their derivatives with carbon monoxide and water yields unsaturated carboxylic acids or their derivatives. Thus, acetylene reacts with carbon monoxide and water to give acrylic acid. With $Ni(CO)_4$ catalyst, higher yields are obtained than with cobalt or iron carbonyls [1, 345].

$$HC\equiv CH + CO + H_2O \xrightarrow{\text{catalyst}} H_2C=CH-COOH$$

Contrary to the hydroformylation or carbonylation of olefins, in the carbonylation of acetylenes addition of the carboxyl group occurs, without exception, to one of the C-atoms of the triple bond. Generally, isomerization of the unsaturated bond via intermediate metalorganic complexes or isomerization of the carbon skeleton are not observed.

As mentioned above, the carbonylation of acetylenes may also be effected without pressure with stoichiometric amounts of $Ni(CO)_4$ as CO donor in the presence of aqueous acids. After reaction of the carbon monoxide which was complexed with nickel, the remaining nickel is found as the salt of the acid present.

$$4\,HC\equiv CH + Ni(CO)_4 + 4\,H_2O + 2\,HX \longrightarrow 4\,H_2C=CH-COOH + NiX_2 + H_2$$

As already stated by Reppe and later confirmed by Ohashi [421–423], the hydrogen resulting from the overall equation could never be found. It is apparently consumed through hydrogenation either of acrylic acid to propionic acid or of acetylene to ethylene or ethane.

To a smaller extent the hydrogen is consumed in a number of side reactions, especially at reaction temperatures below 30 °C where hydrogenation of the unsaturated acid hardly occurs.

Jones et al. [367, 425] formulated the stoichiometry of the hydrocarboxylation of acetylene with regard to the side reactions as follows:

$$10\,RC\equiv CR + 2\,Ni(CO)_4 + 5\,H_2O + 4\,HX \longrightarrow 5\,RC(COOH)=CHR + 2\,NiX_2 + [5\,RC\equiv CR, 3\,CO, 4\,H]$$

The compounds in brackets of the above equation represent a complex mixture of by-products.

The described hydrocarboxylation reaction can be applied generally to all acetylenes [1, 345, 349, 426]. Under usual reaction conditions the elements of formic acid (H–COOH) are added nearly exclusively in cis-position and preferably in the Markovnikov sense [421].

Exceptions with an anti-Markovnikov addition are observed in the hydrocarboxylation of octyne [427], nonyne-2 [345], phenylacetylene, 1-phenylpropyne-1 [345] and propargyl alcohol [428].

Table 36 shows some of the reaction products obtained in the hydrocarboxylation of alkynes. The reactivity of the alkyne in the hydrocarboxylation reaction is largely determined by the substituents connected to the C-atoms of the triple bond. Jones *et al.* [429] stated that the substituents of the triple bond can be classified into two groups:

(A) in substituents which favor hydrocarboxylation:

$$\text{alkyl, aryl, } CH_2OH, CH_2O-OC-CH_3, CHR-O-OC-CH_3,$$
$$CH_2-CH_2-OH, CH_2-CH(CH_3)-OH, CH_2-CH_2O-OC-CH_3$$
$$CH_2-CH_2-O-\overset{\lceil\qquad\rceil}{C}HO(CH_2)_4 \quad \text{and} \quad CH_2-CH_2-CH_2OH$$

(B) in substituents which retard hydrocarboxylation:

$$H, CHROH, CR_2OH, CR_2OOC-CH_3, CH_2-C(OH)-(CH_3)_2,$$
$$CO-CH_3, COOH, COOC_2H_5$$

In acetylenes substituted by a substituent of group (A) plus one of group (B), the carboxyl group is added to the C-atom of the triple bond which is substituted by a class (A) substituent.

An ethyne substituted by two substituents of group (B) reacts very slowly. An exception from the above rule is acetylene which is very reactive.

Vinylacetylene reacts slowly. Reaction velocity is sufficient only in the presence of pyridines [369]. The prior-formed carboxydienes may undergo Diels-Alder addition to give unsaturated dicarboxylic acids in a consecutive reaction [345, 369, 428, 433].

$$HC\equiv C-CH=CH_2 + CO/H_2O \longrightarrow \left(\begin{array}{c} H_2C=C-CH=CH_2 \\ | \\ COOH \end{array}\right)$$

$$2\,H_2C=C-CH=CH_2 \longrightarrow$$
$$\qquad\qquad | $$
$$\qquad\quad COOH$$

Attempts to hydrocarboxylate diacetylenes have failed so far [369].

Different results are obtained in the conversion of halogenated acetylenes. Thus, a halogen atom bond to a C-atom of the triple bond is replaced by hydrogen:

$$2\,RC\equiv CI + 2\,Ni(CO)_4 + 2\,AcOH \longrightarrow 2\,RC\equiv CH + Ni(OAc)_2 +$$
$$NiI_2 + 8\,CO$$

Dehalogenation of haloalkyne is slower than hydrocarboxylation. However, carboxylation does not occur until dehalogenation of the halogenated alkyne is completed. Halogen alkynes substituted by halogen atoms

Table 36. *Hydrocarboxylation of alkynes with carbon monoxide and water*

Acetylene	Reaction product	Yield (%)	Ref.
Acetylene	Acrylic acid	95	[345, 427, 428]
Propyne	Methacrylic acid	50	[428]
Butyne-1	1-Pentenoic acid-2	45	[428]
Hexyne-1	1-Heptenoic acid-2	35	[429]
Octyne-1	1-Nonenoic acid-2	20	[345]
Nonyne-2	2-Decenoic acid-2 2-decenoic acid-3 +	32.5	[345]
Decyne-5	5-Undecenoic acid-5	52	[425]
Phenylacetylene	Atropic acid and traces of cinnamic acid	48	[345, 429]
1-Phenylpropyne-1	α-Methyl cinnamic acid + α-phenyl crotonic acid	54	[345]
Diphenylacetylene	α-Phenyl-trans-cinnamic acid	48	[345, 425, 430]
3-Acetoxypropyne	3-Acetoxybutenoic acid-1	32	[429]
3-Hydroxybutyne-1	3-Hydroxypent-1-enoic acid-2	—	[345]
1-Hydroxybutyne-3	α-Methylene-γ-butyrolactone	23	[429]
1-Acetoxybutyne-3	4-Acetoxypent-1-enoic acid-2	32	[429]
4-(2-Hydroxytetra- hydropyranyl)- butyne-1	4-(2-Hydroxytetrahydropyranyl)- pent-1-enoic acid-2	20	[429]
	4-(2-Hydroxytetrahydropyranyl)- pent-1-enoic acid-2-ethyl ester		
1,4-Diacetoxybutyne-2	1,4-Diacetoxypent-2-enoic acid-2	58	[425]
1-Hydroxypentyne-4- p-Toluene sulfonic acid ester	5-Hydroxyhex-2-enoic acid-2- p-Toluene sulfonate	1.5	[369]
2-Hydroxypentyne-4	α-Methylene-γ-methyl-γ-butyro- lactone	30	[429]
2-Hydroxypentyne-3	4-Hydroxyhex-2-enoic acid-2	60	[425]
3-Acetoxyhexyne-1	3-Acetoxyhept-1-enoic acid-2	48	[429]
2-Hydroxy-2-methyl- pentyne-4	α-Methylene-γ-dimethyl-γ-butyro- lactone	7.4	[429]
1-Acetoxy-1-ethyne- cyclohexane	α-(1-Acetoxycyclohexyl)-acrylic acid	3.5	[429]
2-Acetoxy-2-phenyl propyne	3-Acetoxy-3-phenylbut-1-enoic acid-2	50	[429]
1-Hydroxy-1-phenyl- propyne	3-Hydroxy-3-phenylbut-1-enoic acid-2	12	[429]
5-Bromopentyne	5-Bromohex-1-enoic acid-2	40	[369]
Pentyne-3-one-2	cis- and trans-4-Ketohex-2-enoic acid-2	30	[425]

Table 36 (continued)

Acetylene	Reaction product	Yield (%)	Ref.
Octyne-3-one-2	2-Keto-non-3-enoic acid-4	40	[425]
1-Hydroxypentyne-4	α-Methylene-δ-valerolactone	21	[429]
Ethyl-pent-3-ynoate-1	3-Carbethoxy-pent-1-enoic acid-2	28	[431]
Ethyl-hex-4-ynoate-1	4-Carbethoxy-hex-1-enoic acid-2	46	[431]
Ethyl-hept-5-ynoate-1	5-Carbethoxy-hept-1-enoic acid-2	40	[431]
Ethyl-oct-6-ynoate-1	6-Carbethoxy-oct-1-enoate-2	37	[425, 431]
Hex-1-ynoic acid-1	n-Butyl fumaric acid	28	[425]
5-Cyanopentyne-1	5-Cyanohex-2-enoic acid-2	39	[431]

at a C-atom of the triple bond obviously exhibit an inhibition effect on the hydrocarboxylation. On the contrary a halogen atom bond to the adjacent or more distant C-atoms has no inhibition effect [425].

α-Halogenalkynes react in the presence of $Ni(CO)_4$ and aqueous acids in the hydrocarboxylation to give allenic acids, keto acids or substituted maleic anhydride [434, 435]. By variation of the halogen atom each of the three products may be the main product.

$$H_3C-C\equiv C-CH_2X \xrightarrow{CO/H_2O}$$

X = Cl	15%	2%	—
= Br	—	5%	3%
= I	—	1%	14%

As will be discussed later, halogen alkyls can assume the function of the acid, thus, the presence of acids is not necessary. However, if an acid is present, allenic carboxylic acids formed tend to react to lactones

$$R_2C=C=CH-COOH \longrightarrow$$

Allyl halogenides may replace the acid in $Ni(CO)_4$ catalyzed reactions [434, 438, 439, 1020, 1022].

Table 37. *Allenic acids by hydrocarboxylation of α-halogenated alkynes*

Acetylene	Allenic acid[a]	Yield[b] (%)	Ref.
3-Chloropropyne-1	Butadiene-2,3-acid-1	6	[436, 437]
3-Chlorobutyne-1	Pentadiene-2,3-acid-1	10	[436]
1-Chlorobutyne-2	2-Methylbutadiene-2,3-acid-1	15	[435]
3-Chloro-3-methyl-butyne-1	4-Methylpentadiene-2,3-acid-1	45	[436]
3-Chloroheptyne-2	2-Butylbutadiene-2,3-acid-1	15	[435, 436]
1-Bromoheptyne-2	2-Butylbutadiene-2,3-acid-1	11.5	[435]
1-Iodoheptyne-2	2-Butylbutadiene-2,3-acid-1	22	[435]
p-Toluenesulfonic acid-heptyne-2-ol-1-ester	2-Butylbutadiene-2,3-acid-1	31	[435]
2-Chloro-2-methyl-octyne-3	2-Butyl-4-methylpentadiene-2,3-acid-1	13	[436]
3-Chloro-3-phenyl-propyne-1	4-Phenylbutadiene-2,3-acid-1	12	[436]

[a] The acid was obtained partly in form of the ethyl ester
[b] Yield of acid + ester

In analogy to equations (5), (6), (7) and (8) on page 80 the following reaction course can be proposed:

$$H_2C=CH-CH_2X + Ni(CO)_4 \longrightarrow H_2C=CH-CH_2-Ni(CO)_2X + 2\,CO$$

$$\xrightarrow{HC\equiv CH} CH_2=CH-CH_2-CH=CH-Ni(CO)_2X \xrightarrow{CO}$$

$$CH_2=CH-CH_2-CH=CH-\underset{\underset{O}{\|}}{C}-Ni(CO)_2X \xrightarrow[2\,CO]{H_2O}$$

$$CH_2=CH-CH_2-CH=CH-\underset{\underset{O}{\|}}{C}-OH + Ni(CO)_4 + HX$$

A number of other allyl halogenide compounds [438] reacts analogously to allyl chloride, the yields being in the range of 70–80 %.

As to the most favorable reaction conditions see reference 1020.

The reaction failed with allyl chlorides substituted by electronegative substituents at the double bond. Thus, the following compounds

$$ClCH_2-CH=CH-C\equiv N, \quad ClCH_2-CH=CH-COOCH_3, \quad ClCH_2-CH=CH-Cl$$

could not be reacted in the described manner [438].

Table 38 [443]. *2-cis-5-Dienenoic acid methyl esters* $R''CH=CH-COOCH_3$
obtained at 20 °C in methanol following the equation

$$R'X + HC\equiv CH + CO + CH_3OH \xrightarrow{Ni(CO)_4} R'-CH=CH-COOCH_3 + HX$$

R″	b.p. (°C/torr)	Yield (%)
$CH_2=CH-CH_2-$	62–63/28	70
$CH_3-CH=CH-CH_2-$	73–75/20	81
$CH_2=C(CH_3)CH_2-$	62–63/16	80
$CH_3-CH=CH-CH(CH_3)-$	75–78/20	40
$CH_3-C(CH_3)=CH-CH_2-$	85–86/12	35
$CH_3-(CH_2)_2-CH=CH-CH_2-$	69–70/2	62
$CH_3-(CH_2)_4-CH=CH-CH_2-$	68–70/0.4	73
$CH_3-(CH_2)_6-CH=CH-CH_2-$	103–104/1	64
$CH_3-(CH_2)_{14}-CH=CH-CH_2-$	150–156/0.2	48
$CH_3-C(CH_3)_2-CH_2-CH=CH-CH_2-$	97–100/8	55
$H_3CCOO-CH=CH-CH_2-$	78–81/1	40
$H_3CCOO-CH_2-CH=CH-CH_2-$	126–130/6	32
$H_3CCOO-CH=CH-CH_2-CH=CH-CH_2-$	140–141/5	80
$H_3CCOO-CH=C(CH_3)-CH_2-$	81–82/1	45
$H_3CCOO-CH_2-CH=CH-CH_2-$	115–117/5	35
$NC-CH_2-CH=CH-CH_2-$	133–134/7	79
2-cyclopentene-1-yl	92–96/12	50
2-cyclohexene-1-yl	105–108/12	50
$C_6H_5-CH=CH-CH_2-$	155–160/5	54
$H_3CCOO-CH=CH-CH_2-CH=CH-CH_2-CH_2-CH=CH-CH_2-$		
	150–155/0.3	19

The reaction products are in all cases cis-substituted conjugated unsaturated esters. Often phenols are formed as by-products, especially when nonpolar solvents are used. This side reaction will be discussed in detail in the section on ring closure reactions with carbon monoxide.

In analogy to allyl halogenides, iodobenzene and other aromatic iodo derivatives can be reacted with $Ni(CO)_4$ and acetylene at reaction temperatures above 100 °C with formation of γ-ketoacids or their esters [418, 442], which may be considered as hydrolysis products of β,γ-unsaturated γ-lactones or as hydrogenation products of α,β-unsaturated γ-ketoacids or -esters. α,β-Unsaturated γ-ketoacids or -esters will be hydrogenated under the reaction conditions, but do not take up carbon monoxide because of the presence of electrophilic substituents. Besides allyl halogenides, allyl alcohols, -ethers and -esters may also be reacted to give unsaturated acids or esters.

$$RCH=CH-CH_2OH + HC\equiv CH + CO \xrightarrow[HCl]{Ni(CO)_4}$$

$$RCH=CH-CH_2-CH=CH-COOH$$

($RCH=CH-CH_2-CH=CH-COOCH_2-CH=CHR$ as by-product in water/acetone)

In this case a deficit of hydrogen chloride below the amount necessary for formation of allyl chloride is applied. Usually higher yields (80–85 %) are obtained than with allyl chlorides, where $NiCl_2$, formed in a side reaction, hinders the carboxylation.

In the case of allyl esters, reaction can also be achieved in the absence of hydrogen chloride if higher temperatures ($\approx$ 130 °C) are applied, at which the esters and nickelcarbonyl react with formation of allyl groups, carbon monoxide and nickel salts [750].

As described by H. W. Sternberg [440], hydrocarboxylation of acetylenes is possible also in alkaline medium, where $(Ni_3(CO)_8)^-$ is believed to function as the CO-donor. Thus, Sternberg obtained 25 % of trans-α-phenyl cinnamic acid besides 67 % of tetraphenyl butadiene, starting from diphenyl acetylene. Starting with octynes J. M. J. Tetteroo reported a considerably lower yield [146]. As mentioned on page 83, different reaction products are obtained with Co- or Fe-carbonyls on the one hand and $Ni(CO)_4$ on the other hand. Contrary to nickelcarbonyl, cobaltcarbonyls are of such activity that the initially formed unsaturated acids are hydrocarboxylated a second time at the double bond. Thus, dicarboxylic acids or their derivatives are obtained by hydrocarboxylation of acetylenes with cobaltcarbonyls as catalysts [226, 388–391, 393–397, 441] (see also table 39).

Table 39 [146]. *Hydrocarboxylation of acetylenes with cobalt catalysts*

	Yield (%) based on acetylene
Methyl acrylate *	3.6–7.3
Cyclopentanone	Traces
Cyclopentenone	up to 3
Dimethyl fumarate	up to 2.5
Dimethyl succinate	13–14
Ethane tricarboxylic acid methyl ester	4.5–6
Dimethyl-γ-keto-pimelate	5.5–8.5

* Reaction at 90–110 °C and 200–300 atm with methanol as solvent. Instead of acids the corresponding esters are obtained (see following chapter)

Surprisingly, acetylenes, carbon monoxide and water react with iron-carbonyls in alcohol solution, especially in alkaline medium, to give hydroquinone in 20–30 % yield [239, 400, 444, 445], besides the expected acrylic acid.

$$Fe(CO)_5 + 4 C_2H_2 + 2 H_2O \xrightarrow{\ OH^\ominus\ } 2 HO-\langle\ \rangle-HO + FeCO_3$$

resp.

$$H_2Fe(CO)_4 + 4 C_2H_2 + 2 H_2O \longrightarrow 2 HO-\langle\ \rangle-OH + Fe(OH)_2$$

Formation of hydroquinone also was observed (in small quantities) in the cobalthydrocarbonyl catalyzed reaction of acetylene and some substituted acetylenes in the presence of phosphoric acid [239].

6.2. Acetylene and Substituted Acetylenes in the Presence of Alcohols [382]

Conjugated unsaturated carboxylic acid esters can be obtained by hydrocarboxylation of acetylenes in the presence of alcohols.

However, the reaction does not take the assumed course according to equation (1)

$$HC{\equiv}CH + CO + HOR \xrightarrow{\text{cat.}} H_2C{=}CH{-}COOR \qquad (1)$$

but follows equations (2) and (3)

$$HC{\equiv}CH + CO + H_2O \xrightarrow{\text{cat.}} H_2C{=}CH{-}COOH \qquad (2)$$

$$H_2C{=}CH{-}COOH + HOR \xrightarrow{\text{cat.}} H_2C{=}CH{-}COOR + H_2O \qquad (3)$$

The first step of the reaction relates to the hydrocarboxylation and the second step to the esterification of the primary acid formed with the alcohol present in the reaction mixture.

As could be proved, carbonylation failed in absolute anhydrous medium under the usual reaction conditions [146, 367]. Reaction proceeds only at high temperatures after formation of water originating from the reaction of the alcohol towards ethers [146, 367].

When water is charged to the reactants in small quantities, the reaction rate only depends on the esterification [146].

Another important factor for a successful formation of esters is the strength of the acid used. Thus, the reaction of alkynes with carbon monoxide and water /HOR failed nearly completely with acetic acid (P_k 4.7) or pivalic acid (P_k 5). The reaction products obtained are exclusively the unsaturated acids. Even the strength of chloroacetic acid (P_k 2.8) is unsufficient; equal amounts of acids and esters are formed. However, with hydrochloric acid ($P_k > 0$) high yields of unsaturated esters are obtained [367].

Principally the described ester synthesis can be achieved with the same metal carbonyl catalysts as used in the hydrocarboxylation. Recently the catalyst systems Pd/HCl, $PdCl_2$ [446] and Pd/HI [447] have been reported to be very effective. These catalysts are suitable for converting the less reactive acetylene carboxylic acid esters. In analogy to cobalt catalysts, Pd causes a double hydrocarboxylation of the alkynes:

$$HC\equiv CH-COOC_2H_5 + CO/HOC_2H_5 \longrightarrow$$
$$C_2H_5OOC-CH=CH-COOC_2H_5 + (C_2H_5OOC)_2-CH-CH_2-COOC_2H_5$$

as by-products could be isolated:

$$(C_2H_5OOC)_2-C=CH-COOC_2H_5 \text{ and } (C_2H_5OOC)_2-C=CH-CH=C-(COOC_2H_5)_2$$

Acetylene dicarboxylic acid esters react analogously [446]. Methyl buty-nene reacts at 80 °C and a pressure of 700 atm in the presence of bis-tri-phenylphosphine palladium-(II)-chloride to give 1.2-dimethyl propene-(1)-dicarboxylic acid-(1,3) diethyl ester [448].

$$HC\equiv C-\underset{\underset{CH_3}{|}}{C}=CH_2 + 2\,CO + 2\,HOR \xrightarrow{(P(C_6H_5)_3)_2PdCl_2}$$

$$H_5C_2OOC-\underset{\underset{H_3C}{|}}{C}=\underset{\underset{CH_3}{|}}{C}-CH_2-COOC_2H_5$$

Table 40 shows some esters obtained by hydrocarboxylation of alkynes in water/alcohol solution.

Table 40. *Unsaturated carboxylic acid esters by hydrocarboxylation of alkynes in mixtures of water and alcohols*

Alkyne	Alcohol	Ester	Ref.
Acetylene	Methanol	Methyl acrylate	[449]
Acetylene	Ethanol	Ethyl acrylate	[450]
Acetylene	Butanol	Butyl acrylate	[451, 452]
Acetylene	2-Ethylhexanol	2-Ethylhexyl acrylate	[454]
Acetylene	Tetrahydro-furfuryl alcohol	Tetrahydrofurfuryl acrylate	[455]
Acetylene	Glycol	Ethyleneglycol diacrylic acid ester	[455]
Acetylene	Butanediol-1,4	Butanediol monoacrylic acid ester	[455]
Methyl acetylene	Methanol	Methyl methacrylate	[1029]
Methyl acetylene	Ethanol	Ethyl methacrylate	[456]
Hexyl acetylene	Ethanol	Octene-1-carboxylic acid-2 ethyl ester	[457]
Phenyl acetylene	Ethanol	Ethyl α-phenylacrylate	[457]
Methyl phenyl acetylene	Ethanol	Ethyl α-methylcinnamate	[457]
Diphenyl acetylene	Ethanol	α,β-Diphenyl-γ-crotono-lactone + diethyl diphenylmaleinate	[458]

6.3. Acetylene and Substituted Acetylenes in Presence of Carboxylic Acids, Hydrogen Halides, Mercaptans or Amines

Carboxylic acid anhydrides are formed in the carbonylation of acetylenes if the alcohols are replaced by carboxylic acids (1).

Thioesters (2) are obtained starting from thioalcohols or thiophenols and carboxylic acid amides starting from amines (3).

Carboxylic acid halogenides are formed with hydrogen halides (4) [459, 460].

$$HC\equiv CH + CO + HOOCR \xrightarrow{\text{cat.}} H_2C=CH-\underset{\underset{O}{\|}}{C}-O-\underset{\underset{O}{\|}}{C}-R \qquad (1)$$

$$HC\equiv CH + CO + HSR \xrightarrow{\text{cat.}} H_2C=CH-\underset{\underset{O}{\|}}{C}-SR \qquad (2)$$

$$HC\equiv CH + CO + HNR_1R_2 \xrightarrow{\text{cat.}} H_2C=CH-\underset{\underset{O}{\|}}{C}-NR_1R_2 \qquad (3)$$

$$HC\equiv CH + CO + HCl \xrightarrow{\text{cat.}} H_2C=CH-\underset{\underset{O}{\|}}{C}-Cl \qquad (4)$$

Very little work in this regard has yet been done compared to syntheses of unsaturated acids and unsaturated esters. Therefore few experimental data are given in the literature.

W. A. Raczynski [461] reacted acrylic acid, acetylene and $Ni(CO)_4$ in inert solvents at 40–50 °C and obtained acrylic acid anhydride in 88.3% yield. Acrylic acid anhydride also is formed starting from acetylene, carbon monoxide and water [462].

The reaction of alkynes with carbon monoxide and mercaptans requires a $Ni(CO)_4$ excess, as certain amounts of $Ni(CO)_4$ are fixed as nickel sulfide (1,3).

Acetylene reacts with hydrogen chloride and carbon monoxide in presence of $RhCl_2$ as catalyst to give acrylic acid chloride [459] whereas with complex Pd-salts mainly succinic acid dichloride [460] and smaller amounts of β-chloropropionyl chloride are formed. When hydrogen chloride is replaced by phosgene, trans, trans-muconic acid dichloride besides fumaric acid dichloride is obtained [463].

The same products are formed if stoichiometric amounts of $PdCl_2$ are used instead of phosgene; $PdCl_2$ is reduced to elementary palladium [464].

$$\begin{array}{c} CH \\ \| \\ CH \end{array} + 2\,CO + PdCl_2 \longrightarrow \begin{array}{c} ClOC \diagdown \, CH \\ \qquad \| \\ HC \diagup \, COCl \end{array} + Pd$$

$$2\begin{array}{c} CH \\ \| \\ CH \end{array} + 2\,CO + PdCl_2 \longrightarrow \begin{array}{c} ClOC \diagdown \, CH \\ HC \diagup \\ \quad CH \\ HC \diagup \diagdown \, COCl \end{array} + Pd$$

The last-cited reaction takes place without participation of active hydrogen atoms; nevertheless, it deals with the carbonylation leading to carboxylic acid derivatives.

Table 41 shows some unsaturated thioesters, synthesized from acetylene; the yields are up to 77 %.

Table 41 [465]. *Thioesters from alkynes, carbon monoxide and SH compounds*

Alkyne	Thiol	Reaction product
Acetylene	Hydrogen sulfide	Thioacrylic acid
Acetylene	Dodecyl mercaptan	Thiododecyl acrylate
Acetylene	Benzyl mercaptan	Thiobenzyl acrylate
Acetylene	Thiophenol	Thiophenyl acrylate
Acetylene	p-Thiocresol	Thio-p-tolyl acrylate
Acetylene	Thioglycolic acid	S-Acryl thioglycolic acid
Phenylacetylene	Ethylmercaptan	Thioethyl-α-phenylacrylate

Under the usual conditions applied in the thioester syntheses, alkynes react with primary and secondary amines in the catalytic as well as in the stoichiometric procedure to give unsaturated acid amides [465–468]. Also acid amides containing a reactive hydrogen atom at the nitrogen can be reacted [465].

However, the synthesis of unsubstituted acrylamide from acetylene, carbon monoxide and ammonia has failed so far. Polymerization was always observed. Table 42 shows some unsaturated amides obtained from primary and secondary amines or amides resp.

The yields of unsaturated amides are in the range of 40–70% if no polymers are formed.

Table 42 [465]. *Unsaturated carboxylic acid amides by reaction of alkynes with carbon monoxide and amines or amides resp.*

Alkyne	Amine	Reaction product
Acetylene	Ethylamine	N-Ethylacrylamide (dimer)
Acetylene	Butylamine	N-Butylacrylamide (dimer)
Acetylene	Aniline	Acrylanilide
Acetylene	Pyrrolidone	N-Acrylpyrrolidone
Acetylene	Bicyclohexylamine	N-Bicyclohexylacrylamide
Acetylene	Diphenylamine	N-Diphenylacrylamide (polymer)
Acetylene	Urea	N-Acryl urea (polymer)
Phenylacetylene	Aniline	α-Phenylacrylanilide
Acetylene	Acetamide	N-Acetylacrylamide

6.4. Olefins and Functional Derivatives in the Presence of Water

The carbonylation of olefins with CO/H_2O had been investigated by many authors in the USA before Reppe started his work.

However, the reaction proceeded only under drastic conditions (pressure 700 upward to 900 atm) in the presence of mineral acids, BF_3 or metal halogenides. At that time metal carbonyls had been regarded as catalyst poisons. However, Reppe could prove that olefins react with carbon monoxide and water in the presence of metal carbonyls. The reaction products are saturated carboxylic acids. Whereas $Ni(CO)_4$ is the preferred catalyst in the carbonylation of acetylenes, cobalt, rhodium and ruthenium catalysts are equivalent or superior in olefin carbonylation. Also palladium and hydrochloric acid containing catalyst systems are of special activity in hydrocarboxylation [469–471]. Iron has an accelerating effect [472]. Addition of boric acid to Ni or Co catalysts increases the catalyst life and suppresses the formation of insoluble polymer products [473].

Generally in the olefin carbonylation stronger reaction conditions have to be applied than in the carbonylation of acetylenes. In the stoichiometric procedure with $Ni(CO)_4$ as catalyst, 150 °C and CO-pressures of about 50 atm are required and in the catalytic procedure conversion can be realized at 250 °C and a pressure of 200 atm. Palladium-containing catalysts are active already at 80–150 °C [469–471, 1021].

At low temperatures carbonylation of olefins succeeds without pressure with stoichiometric quantities of cobalt hydrocarbonyl, especially in the preparation of esters.

Contrary to acetylene, which does not undergo isomerization of inter-mediate acetylene metal complexes, in the presence of $Ni(CO)_4$ olefins isomerize in many cases. In turn products are obtained with a carboxyl group linked at a C-atom which does not belong to the double bond of the starting material [146], table 43.

Table 43 [146]. *Composition of n-octane carboxylic acids obtained by stoichiometric hydrocarboxylation of n-octenes in presence of $Ni(CO)_4$ as catalyst*

Octene	Octane carboxylic acids (%)			
	–(1)	–(2)	–(3)	–(4)
cis-(4)	–	2.5	4.3	93.2
cis-(1)	37.6	61.0	0.8	0.8

With cobalt carbonyl isomerizations are more favored.

Many experiments have been carried out with the objective of reducing the drastic reaction conditions which are necessary for carbonylation of olefins. Tetteroo [474] succeeded in stoichiometric hydrocarboxylation of olefins at 55–60 °C and atmospheric pressure with UV irradiation. In the presence of cobalt catalysts the reaction is accelerated remarkably by addi-tion of 5–10 % of hydrogen to carbon monoxide (about factor 3). Obviously the acceleration is caused by favoring the formation of hydrocarbonyl from $Co_2(CO)_8$ and hydrogen.

Also, addition of pyridine has the same effect (for details see the next chapter). Generally, olefins react with $Ni(CO)_4$ without admixture of halo-gens or halogen compounds. Only branched olefins like isobutylene require addition of halogens [475, 476].

These olefins react in the stoichiometric procedure as well as in the catalytic method with $Ni(CO)_4$ to give about 10 % of trialkyl acetic acid besides 90 % of acids branched in the β-position [35]. High yields of trialkyl acetic acid are obtained only at low temperatures so far [35].

According to Levering and Glaserbrook [477] the reaction rate of olefin, carbon monoxide and water with addition of iodine is about 40 times that without halogens (table 45).

For a long time experiments failed to produce normal carbonylation products starting from dienes. Under the usual conditions the conjugated dienes underwent a Diels-Alder reaction with subsequent hydrocarboxy-lation of the prior formed isolated alicyclic dienes. Thus, e. g. butadiene first reacted to give vinylcyclohexene which then yielded a mixture of dicarboxylic acids [493]. In other cases cyclic ketones were obtained from dienes and carbon monoxide (see section on ring closure reaction with carbon monoxide).

Table 44 [477]. *Increase of reaction velocity in the hydrocarboxylation of olefins by addition of iodine*

Olefin	Mol	$Ni(OAc)_2$ (mol)	H_2O (mol)	HI (mol)	Temp. (°C)	Pressure (atm)	Reaction time (h)	Conversion of olefin	Space velocity (g mol/ l/h)
Pentene-1	1.5	0.05	2.0	—	300	406–420	8.0	93.5	0.5
	1.5	0.05	2.0	—	300	406–420	8.5	92.8	0.5
	2.9	0.07	2.6	0.013	300	406–420	0.2	80.6	20.5
	2.0	0.07	2.6	0.013	300	406–420	13.0	82.0	0.2
Hexene-1	1.2	0.05	1.6	—	296	406–420	13.0	82.0	0.2
	1.2	0.05	1.6	—	300	406–420	12.0	79.0	0.2
Pentene-2	2.0	0.05	2.5	—	300	406–420	12.5	66.0	0.2
	2.0	0.05	2.5	—	300	406–420	14.5	71.4	0.2
	2.0	0.05	2.5	0.013	300	406–420	0.4	88.0	10.0
	2.0	0.05	2.6	0.013	300	406–420	0.4	88.6	10.0
2-Methyl-butene-2	1.5	0.05	2.0	—	300	406–420	2.0	No reaction	
	1.5	0.05	2.0	—	300	406–420	5.0	No reaction	
	2.0	0.05	2.5	0.013	300	406–420	0.04	81.5	8.1
Cyclohexene	1.7	0.05	2.2	0.013	320	420	0.2	72.2	8.2
Mixtures of pentenes	2.0	0.074	2.3	0.01	320	420	0.6	73.0	4.3

Table 45 [477]. *Variation of straight chain to branched acids ratio in the
hydrocarboxylation of olefins by addition of iodine*

| Olefin | Promotor | Acids (%) | |
		Straight chain	Bran- ched
Pentene-1	—	40.2	59.8
Pentene-1	HI	47.4	52.6
Hexene-1	—	39.9	60.1
Pentene-2	—	28.8	71.2
Pentene-2	HI	44.3	55.7
2-Methylbutene-2	—	—	—
2-Methylbutene-2	HI	—	100

Some products available by hydrocarboxylation of olefins are listed in
table 46.

Recently a direct hydrocarboxylation of butadiene succeeded at moderate temperatures (about 120 °C) with a Pd-containing catalyst system
[469–471]. Formally the reaction proceeds in the same manner as hydro-
chlorination; by 1.4 addition of the formic acid elements, butene-2-carboxy-
lic acid-1 is formed.

Imyanitov succeeded in direct dicarboxylation of conjugated diolefins
[494, 495, 1025–1027] with cobalt hydrocarbonyl in the presence of pyridine
at 200 to 210 °C and a pressure of 250 atm. Other N-bases are less effective.

The authors proved that first an 1.4-addition of the hydrocarbonyl to
the dienes (1) takes place, these being remarkably more reactive than monoolefins or isolated dienes. Subsequently pyridine reacts with the unsaturated
acyl cobalt tetracarbonyl to form a salt, as follows (2).

(1) $H_2C=CH-CH=CH_2$ + $HCo(CO)_4$ $\xrightarrow{CO}$ $H_3C-CH=CH-CH_2-\underset{\underset{O}{\|}}{C}-Co(CO)_4$

(2) $H_3C-CH=CH-CH_2-\underset{\underset{O}{\|}}{C}-Co(CO)_4$ + $N\diagdown\bigcirc$ $\longrightarrow$

$\left(CH_3-CH=CH-CH_2-\underset{\underset{O}{\|}}{C}-N\bigcirc\right)^{\oplus}\left(Co(CO)_4\right)^{\ominus}$

The formation of this salt inhibits the normally expected formation of
cyclic ketones from dienes and carbon monoxide. Under more drastic conditions the complex formulated in (2) reacts subsequently with H_2O via a

Table 46. *Hydrocarboxylation of olefins*

Olefin	Catalyst	Products	Yield (%)	Ref.
Ethylene	NiCl$_2$, Co Pd(P(C$_6$H$_5$)$_3$/HCl Pd/HI/I$_2$	Propionic acid	90	[470, 471, 478– 483, 490]
Propene	Ni(CO)$_4$	Butyric acid + isobutyric acid (2:1)	60	[478, 480, 483]
Propene	Co$_2$(CO)$_8$	Butyric acid	88	[484]
Butene-1	Ni(CO)$_4$	n- and iso-Valeric acid	—	[483]
Isobutene	Ni(CO)$_4$/NiI$_2$	Isovaleric acid + trimethyl acetic acid (6:1)	100	[478, 483]
Hexene-1		2-Methylhexanoic acid + heptanoic acid	70	[485]
Octene-1	Ni/silica gel	2-Methyloctanoic acid + nonanoic acids	84	[483, 486]
2-Ethylhexene-1	Raney-Ni, NiI$_2$, CuI	2-Ethylheptanoic acids	60	[483]
2-Ethylhexene-1	Co$_2$(CO)$_8$	Nonanoic acids	57	[484]
1-Vinylcyclo- hexene-3	Pd(P(C$_6$H$_5$)$_3$)$_2$Cl$_2$/HCl	Isomeric (carboxy- cyclohexyl)- propionic acids		[469]
Dodecene-1	Ni/silica gel	2-Methyldodecanoic acid + tridecanoic acid	28	[483]
Octadecene-1	Ni(CO)$_4$	2-Methyloctadecanoic acid	67	[483]
Cyclohexene	Ni(CO)$_4$	Cyclohexane carboxylic acid	78	[482, 483, 485, 487]
Cyclohexene	Co$_2$(CO)$_8$	Cyclohexane carboxylic acid	89	[484]
Cyclooctene	Ni(CO)$_4$	Cyclooctane carboxylic acid	31	[488]
Cycloocta- diene-1,5	Pd(P(C$_6$H$_5$)$_3$)$_2$Cl$_2$/HCl	Cyclooctene-4-carboxy- lic acid-1	51	[469]
Cycloocta- diene-1,5	Pd(P(C$_6$H$_5$)$_3$)$_2$Cl$_2$/HCl	Cyclooctane dicarboxy- lic acid-1,5		[469]

Table 46 (continued)

Olefin	Catalyst	Products	Yield (%)	Ref.
Bicyclo-(2.2.1)-heptene		Bicyclo-(2.2.1)-heptane-2 carboxylic acid	80	[489]
cis, trans, trans-Cyclododeca-triene-1,5,9	$Pd(P(C_6H_5)_3)_2Cl_2$	Cyclododecadiene monocarboxylic acid		[469]
Bicyclo-(2.2.1)-heptadiene	$Pd(P(C_6H_5)_3)_2Cl_2/HCl$	Bicyclo-(2.2.1)-heptene-5-carboxylic acid	80	[489]
Cyclododeca-triene-1,5,9	$PdI_2/HI/I_2$	Cyclododeca-5,9-dicarboxylic acid + cyclododecene-9-dicarboxylic acid-1,5		[471]
Cyclooctadiene		Cyclooctane carboxylic acid		[491]
Hexadiene-1,5	$Ni(CO)_4$	2-Methylhexen-5-oic-acid-1	20	[482, 483]
Hexadiene-1,5	$Co_2(CO)_8$/pyridine	Suberic acid	39	
Cyclododeca-triene	$Co/CoSO_4$ (5:1)	Cyclododecane carboxylic acid + cyclododecadiene carboxylic acid	16	[492]
Butadiene-1,3	$(Pd(P(C_6H_5)_3)_2Cl_2/HCl$	Butene-2-carboxylic acid-1	70	[469]
Butadiene-1,3	$Pd/HCl/O_2$	Butene-2-carboxylic acid-1		[471]
Butadiene-1,3	$Ni(CO)_4$	2-(Carboxycyclohexyl)-propionic acid		[482]
Butadiene-1,3	$Pd(P(C_6H_5)_3/HCl$	Butene-2-carboxylic acid-1		[470]
Undecylenic acid	$Ni(CO)_4$	Dodecane-1,12-diacid	54	[483, 485]
Undecylenic acid	$Co_2(CO)_8$	Dodecane diacid	76	[484]
Butene-3-one-1	$Ni(CO)_4$	Levulinic acid + 2-methyl acetoacetic acid	20	[483]
Dihydrofuran	$Ni(CO)_4$	Tetrahydrofuran carboxylic acid		[483]
Styrene	$Pd(P(C_6H_5)_3/HCl$	α-Phenyl propionic acid		[470]

diacyl compound to give dicarboxylic acid. Thus, butadiene reacts to give a mixture of adipic acid, methyl glutaric acid and methyl succinic acid. It can be assumed that the formation of the isomeric acids is based on formation of intermediate π-complexes as observed in the hydroformylation of unsaturated esters.

Instead of dicarboxylic acids the unsaturated monocarboxylic acids can be obtained, e.g. if the salt formed in the first stage is allowed to react with water [494, 495].

$$H_2C=CH-CH=CH_2 \xrightarrow[HCo(CO)_4/Py]{CO} \left[H_3C-CH=CH-CH_2-\overset{\displaystyle \underset{\|}{O}}{C}-N\bigcirc \right]^{\oplus} Co(CO)_4^{\ominus}$$

$$\swarrow CO/HCo(CO)_4Py/H_2O \qquad\qquad H_2O \searrow$$

$$HOOC-(CH_2)_4-COOH \qquad\qquad H_3C-CH=CH-CH_2-\overset{\displaystyle \underset{\|}{O}}{C}-OH$$

$$HOOC-\underset{\underset{\displaystyle CH_3}{|}}{CH}-CH_2-CH_2-COOH$$

$$HOOC-\underset{\underset{\displaystyle C_2H_5}{|}}{CH}-CH_2-COOH$$

With rhodium catalysts in the absence of pyridine conjugated dienes can be converted into unsaturated esters [496].

The carbonylation of diolefins with isolated double bonds can be realized in the same manner as with monoolefins and results in dicarboxylic acids.

The selective hydrocarboxylation of one double bond in e.g. cyclooctadiene-1,5 or cyclododecatriene-1,5,9, succeeds with the highly reactive complex Pd-catalysts whereas the other double bonds in the molecule are unchanged [469]. In the presence of Ru-catalysts allene reacts to give methacrylic acid [497, 498].

$$H_2C=C=CH_2 \xrightarrow[Ru_2(CO)_9]{CO/H_2O} H_2C=\underset{\underset{\displaystyle CH_3}{|}}{C}-COOH$$

Unsaturated carboxylic acids and unsaturated alcohols undergo the expected reactions with nickel or cobalt catalysts and form dicarboxylic acids or hydroxy carboxylic acids or their lactones [499–502].

However, the last mentioned reaction may also proceed according to the mechanism of ring closure reactions with carbon monoxide as described in the 4th section of this monograph. It will be discussed in detail later.

$$H_2C=CH(CH_2)_8COOH \xrightarrow{CO/H_2O}$$

$$HOOC-(CH_2)_{10}-COOH \; + \; HOOC-\underset{\underset{CH_3}{|}}{CH}-(CH_2)_8-COOH$$

$$H_3C-(CH_2)_7-CH=CH-(CH_2)_7-COOH \xrightarrow[Ni/I^{\ominus}]{CO/H_2O}$$

$$H_3C-(CH_2)_7-\underset{\underset{COOH}{|}}{CH}-(CH_2)_8-COOH \; + \; H_3C-(CH_2)_8-\underset{\underset{COOH}{|}}{CH}-(CH_2)_7-COOH$$

$$H_2C=CH-CH_2-CH_2-OH \xrightarrow{CO/H_2O}$$

Reppe recommended as catalysts for the carbonylation of α,β-unsaturated acids $K_2(Ni(CN)_4)$ and $K_2(Ni(CN)_3)$ in the presence of alkali carbonate [503].

$$H_2C=CH-COONa \xrightarrow[150\,°C/200\,atm]{CO,\ alkali\ carbonate} NaOOC-CH_2-CH_2-COONa$$

The formation of five and six-ring ketones on carbonylation of 1,5-dienes was reported by Asinger et al. [1018].

6.5. Olefins and Functional Derivatives in the Presence of Alcohols

The reaction of olefins and their functional derivatives with carbon monoxide and alcohols to saturated carboxylic acid esters generally proceeds at a lower velocity than the formation of the free acids as illustrated in the last chapter [504]. In the presence of nickel halogenide catalysts, reaction temperatures between 180–200 °C and pressures from 100 to 200 atm are required. Yields are in the range of 90%. With cobalt catalysts reaction temperatures between 140 to 170 °C are recommended [505].

Reaction time has to be as short as possible in order to suppress side reactions. Especially the conversion of alcohols to carboxylic acids with one more C-atom, which will be discussed subsequently, or formation of ethers and free acids may be undesirable reactions [504]. The carbonylation rate of olefins with carbon monoxide/ROH in the presence of Co carbonyls can be accelerated remarkably by addition of small portions of hydrogen to carbon monoxide, which favors the formation of hydrocarbonyl. The same effect can be achieved by addition of pyridine.

By addition of pyridine as well as hydrogen, carbonylation of olefins proceeds with nearly the same reaction velocity as observed in hydroformylation reactions [506, 507].

Besides nickel and cobalt, almost all of the catalysts discussed in the last chapter which were suited for the formation of free acids can be applied, e.g. rhodium, palladium and, with certain restrictions, iron. Cobalt hydro-carbonyl catalyzes the stoichiometric ester synthesis at mild reaction conditions [35, 121]. The initially formed acylcobalt carbonyls react rapidly with alcohols even at 50 °C and, in the presence of Na-alcoholate, even at 0 °C to give esters [121]. Dienes with isolated double bonds react with carbon monoxide and alcohols at mild reaction conditions in the presence of Pd/HCl to give unsaturated monocarboxylic acid esters and at more severe conditions to give saturated dicarboxylic acid esters [508].

Also trienes such as cyclododecatriene-1,5,9 yield mono- and dicarboxylic acid esters [488, 509]. Cyclododecane tricarboxylic acid esters are formed with bis-triphenyl phosphine palladium dichloride as catalyst [448]. The catalyst can be recycled [517]. Nearly quantitative syntheses of monocarboxylic acid esters of cyclododecadiene and of tricarboxylic acid esters of cyclododecane can be achieved with complex Pd-catalysts.

In the presence of Ni or Co catalysts, only saturated monocarboxylic acids could be obtained starting from cyclooctadiene [491] as well as from cyclododecatriene [492]. Dienes with conjugated double bonds may be converted with palladium catalysts [448, 511]. Butadiene reacts with carbon monoxide and ethanol at 70 °C in the presence of $PdCl_2$ or di-benzonitrile palladium dichloride to give the ethyl ester of penten-3-oic-acid-1:

$$H_2C=CH-CH=CH_2 \xrightarrow[CO/C_2H_5OH]{PdCl_2} H_3C-CH=CH-CH_2-COOC_2H_5$$

whereas from isoprene a mixture of ethyl 4-methylpent-3-enonate-1 and the ethanol adduct ethyl 4-ethoxy-4-methylpentanoate, besides some other products, is obtained.

Allene reacts in presence of $PtCl_2/ZnCl_2$ [498] as well as $Ru_2(CO)_9$ [512] with carbon monoxide and alcohols to give methacrylic acid esters. Yields obtained are between 40 and 50 % [497].

$$H_2C=C=CH_2 \xrightarrow{CO/HOR} \underset{\underset{CH_3}{|}}{CH_2=C-COOR}$$

Reaction of monoolefins, $PdCl_2$ and carbon monoxide in nonpolar solvents, with subsequent treatment with alcohols, yields chlorocarboxylic acid esters [513, 514]. In this case $PdCl_2$ has to be present in stoichiometric quantities as a Cl source.

$$H_2C{=}CH_2 \xrightarrow[-Pd]{PdCl_2/CO} Cl{-}CH_2{-}CH_2{-}COCl \xrightarrow[-HCl]{+HOR} Cl{-}CH_2{-}CH_2{-}COOR$$

The intermediate β-chloropropionyl chloride could be isolated and identified as the anilide [515].

Besides olefins the highly reactive cyclopropanes also can be reacted. Thus, cyclopropane and propylene yield the same reaction products [516]. Unsaturated esters can be carbonylated analogously to olefins [35, 121, 448] (see also table 48).

Table 47 [514]. *Chlorocarboxylic acid esters by carboxylation of monoolefins in presence of $PdCl_2$*

Olefin	Alcohol	Products	Yield (%)
Ethylene	Ethanol	$Cl{-}CH_2{-}CH_2{-}COOC_2H_5$	41
Propylene	Methanol	$H_3C{-}CHCl{-}CH_2{-}COOCH_3$	27
Butene-1	Methanol	$H_3C{-}CH_2{-}CHCl{-}CH_2{-}COOCH_3$	11
Pentene-1	Methanol	$H_3C{-}CH_2{-}CH_2{-}CHCl{-}CH_2{-}COOCH_3$	10
Butene-2	Methanol	$H_3C{-}CHCl{-}CH(CH_3){-}COOCH_3$	13
Isobutene	Methanol	$H_3C{-}C(CH_3)(Cl){-}CH_2{-}COOCH_3$	12
Cyclohexene	Methanol	Methyl 2-chloro-cyclohexane carboxylate-1	36
Allyl chloride	Methanol	$Cl{-}CH_2{-}CHCl{-}CH_2{-}COOCH_3$	5
Hexene-1	Methanol	$C_4H_2{-}CHCl{-}CH_2{-}COOCH_3$	9
Isobutene	Ethanol	$(CH_3)_9{-}CCl{-}CH_2{-}COOC_2H_5$	
Vinyl chloride	Methanol	$Cl_2CH{-}CH_2{-}COOCH_3$	5

In the presence of bis-triphenylphosphine palladium-II-chloride as catalyst, the carbonylation of Diels-Alder products can be achieved even at 40 °C, e.g. with ethyl tetrahydrophthalate or diethyl bicyclo- (2.2.1)-heptene-(4)-dicarboxylate-1,2, which have a tendency to cleave into the starting materials. The catalyst combination $(P(C_2H_5)_3)_2PdCl_2$ plus piperidine-triphenylphosphine palladium dichloride allows the carbonylation of phenyl-vinylsulfonate [406]. The relatively expensive but easily available catalyst [518] can be recycled many times.

Table 48. *Carboxylation of olefins in the presence of alcohols*

Olefin	Alcohol	Catalyst	Products	Yield (%)	Ref.
Ethylene	Methanol	$PtCl_2/SnCl_2 \cdot 2H_2O/$ HCl	Methyl propionate	72	[512]
Ethylene	Methanol	$CoI_2/NiI_2/FeI_2$	Methyl propionate		[519]
Ethylene	Methanol	Co propionate	Methyl propionate		[520]
Ethylene	Ethanol	$PdCl_2/HCl$ or Pd/HCl	Ethyl propionate		[522]
Ethylene	Ethanol	$Pd(P(C_6H_5)_3)_2Cl_2$	Ethyl propionate	90	[406, 1021]
Ethylene	Ethanol	$PtCl_2/SnCl_2 \cdot 2H_2O/$ HCl	Ethyl propionate	85	[512]
Ethylene	Ethanol	Co propionate	Ethyl propionate		[520]
Ethylene	Isopropanol	$PtCl_2/SnCl_2 \cdot 2H_2O/$ HCl	Isopropyl propionate	46	[512]
Ethylene	Butanol	$Ni(CO)_4$	Butyl propionate	72	[521]
Propylene	Methanol	Co propionate	Methyl butyrate		[520]
Propylene	Methanol	$Co_2(CO)_2$	Methyl butyrate + methyl isobutyrate	85	[523]
Propylene	Ethanol	$PdCl_2/HCl$ or Pd/HCl	Ethyl butyrate + ethyl isobutyrate		[522]
Propylene	Dodecanol	$Pd(P(C_6H_5)_3)_2Cl_2$	Dodecyl butyrate	95	[448]
Propylene	Glycol	$Pd(P(C_6H_5)_3)_2Cl_2$	Glycol butyrate + glycol isobutyrate	15	[448]
Propylene	Phenol	$Pd(P(C_6H_5)_3)_2Cl_2$	Phenyl butyrate	68	[448]
Butene-2	Methanol	$PtCl_2/SnCl_2 \cdot 2H_2O/$ HCl	Methyl valerate		[512]
Butene-1	Methanol	Co propionate	Methyl valerate		[520]
Butene-2	Methanol	$Pd(P(C_6H_5)_3)_2Cl_2$	Methyl 2-methyl-butyrate	95	[520, 1021]
Isobutene	Ethanol	Co propionate	Methyl isovalerate		[520]
Isobutene	Ethanol	$Pd(P(C_2H_5)_3)_2Cl_2$	Ethyl pivalate		[448]
Octene-1	Ethanol	NiI_2	Ethyl isopelargonate	95	[524]
Octene-1	Phenol	$Ni(CO)_4$	Phenyl nonanoate + phenyl nonanoate-2	26	[525]
Isooctene	Ethanol	$Pd(P(C_6H_5)_3)_2Cl_2$	Ethyl nonanoate (isomers)		[448]
Cetene-1	Ethanol	Ni/NiI_2	Mixture of ethyl undecanoates	61	[525]

Table 48 (continued)

Olefin	Alcohol	Catalyst	Products	Yield (%)	Ref.
Cyclo-hexene	Methanol	Cobalt	Methyl hexa-hydrobenzoate	76–86	[526]
Cyclo-hexene	Methanol	$Co_2(CO)_8$	Methyl hexa-hydrobenzoate	82	[522]
Cyclo-hexene	Methanol	$PtCl_2/SnCl_2 \cdot 2H_2O/$ HCl	Methyl hexa-hydrobenzoate	1	[512]
Octa-decene-1	Ethanol	NiI_2	Ethyl nonade-canoate	65	[525]
Octa-decene	Ethanol	Ni/CuI_2 on silica gel	Ethyl α-methyl-stearate	75	[525]
Allene	Methanol	$PtCl_2/SnCl_2 \cdot 2H_2O$	Methyl metha-crylate	46	[512]
Butadiene	Methanol	$(P(C_4H_9)_3PdCl_2)_2$	Methyl pent-2-enoate-1	71	[527, 1021]
Butadiene	Ethanol	$PdCl_2/HCl$	Ethyl pent-2-enoate-1		[511]
Butadiene	Ethanol	$(C_6H_5CN)_2PdCl_2/$ HCl	Ethyl pent-2-enoate-1	72	[448]
Butadiene	Ethanol	$PtCl_2/SnCl_2 \cdot 2H_2O$	Ethyl pent-2-enoate-1		[512]
Isoprene	Methanol	$(P(C_4H_9)_3PdCl_2)_2$	Methyl 4-methyl-pent-3-enoate	38	[527]
Penta-diene-1,3	Methanol	$(P(C_4H_2)_9PdCl_2)_2$	Methyl 2-methyl-pent-3-enoate	34	[527]
Hexa-diene-1,5	Methanol	$Pd(P(C_4H_2)_3)_2I_2$	Methyl 2-keto-3-methylcyclo-pentylacetate	50	[527]
2,3-Di-methyl-buta-diene-1,3	Methanol	$(P(C_4H_9)_3PdCl_2)_2$	Methyl 3,4-dime-thylpent-3-enoate	50	[527]
3-Methyl-hepta-triene-1,4,6	Ethanol	$Pd(P(C_6H_5)_3)_2Cl_2/$ HCl	Ethyl 3-methyl-heptadienoate	28	[448]
1-Vinyl-cyclo-hexene-3	Methanol	$Pd(P(C_6H_5)_3)_2Cl_2/$ HCl	Methyl α-(cyclo-hexene-3-yl-1)-propionate	85	[448, 1021]
Styrene	Ethanol	$Pd(P(C_6H_5)_3)_2Cl_2/$ HCl	Ethyl α-phenyl-propionate	36	[448]
p-Diiso-propenyl-benzene	Ethanol	$Pd(P(C_6H_5)_3)_2Cl_2/$ HCl	Diethyl 2-phenyl-enedi-2-isobutyrate	47	[448]

Table 48 (continued)

Olefin	Alcohol	Catalyst	Products	Yield (%)	Ref.
Cyclo-octene	Methanol	$Pd(P(C_6H_5)_3)_2Cl_2$	Methyl cyclo-octane-carboxylate	50	[448]
Cyclo-dode-cene	Ethanol	$PdCl_2/HCl$	Ethyl cyclo-dodecane-carboxylate	93	[509]
Cyclo-penta-diene	Methanol	$(P(C_4H_9)_3PdCl_2)_2$	Methyl cyclo-pent-2-ene-carboxylate	73	[527]
Cycloocta-diene-1,3	Methanol	$(P(C_4H_9)_3)_2PdI_2$	Methyl cyclo-octene-2-carboxylate	14	[528]
Cycloocta-diene-1,3	Ethanol	$PdCl_2/HCl$	Ethyl cyclo-octene-2-carboxylate	19	[529]
Cycloocta-diene-1,5	Methanol	$Pd(P(C_4H_9)_3)_2I_2$	Methyl cyclo-octene-4-carboxylate + dimethyl cyclo-octane-dicarboxylate	45 / 30	[528]
Cycloocta-diene-1,5	Ethanol	Pd-acetylacetonate	Ethyl cyclo-octene-4-carboxylate	75	[529]
Cycloocta-diene-1,5	Ethanol	PdI_2	Diethyl cyclo-octane-dicarb-oxylate	95	[529]
Cycloocta-diene-1,5	Chloro-ethanol	$Pd(P(C_6H_5)_3)_2Cl_2$	Ethyl 2-chloro cyclooctene-4-carboxylate	37	[448]
1-Phenyl-deca-triene-1,4,8	Methanol	$Pd(P(C_6H_5)_5)_2Cl_2$	Methyl 1-phenyl-decadiene-carboxylate		[448]
Cyclo-dodeca-triene-1,5,9	Methanol	$(P(C_6H_5)_3)_2PdBr_2$	Methyl cyclo-dodeca-5,9-dienecarboxy-late-1 + dimethyl cyclo-dodecene-9-dicarb-oxylate-1,5	40 / 10	[448, 512]

Table 48 (continued)

Olefin	Alcohol	Catalyst	Products	Yield (%)	Ref.
Cyclo-dodeca-triene	Ethanol	PdCl$_2$/HCl	Ethyl cyclo-dodeca-5,9-diene-carboxylate diethyl cyclo-dodecene-9-dicarboxylate	91	[509]
Cyclo-dodeca-triene	Ethanol	(Pd(P(C$_6$H$_5$)$_3$)$_2$Cl$_2$/ HCl	Triethyl cyclo-dodecanetri-carboxylate	70–80	[448, 1021]
Methyl-vinyl-acetate	Methanol	(P(C$_6$H$_5$)$_3$)$_2$PdCl$_2$/ HCl	Dimethyl methyl-succinate	67	[448]
Ethyl-unde-cyle-noate	Ethanol	Ni(CO)$_4$/CuI	Diethyl dodecanoate	78	[525]
Methyl-oleate	Methanol	Ni(CO)$_4$	Dimethyl hepta-decane-dicarboxylate	36	[525]
Diethyl-tetra-hydro-phtha-late	Ethanol	(P(C$_6$H$_5$)$_3$)$_2$PdCl$_2$/ HCl	Triethyl perhydro-trimellitate	86	[448]
Diethyl-bicyclo-(2.2.1)-heptene-4-dicarb-oxy-late-1,2	Ethanol	(P(C$_6$H$_5$)$_3$)$_2$PdCl$_2$/ HCl	Triethyl bicyclo-heptane-2,2,1-tricarboxylate-1,2,4	52	[448]
Vinyl-chloride	Ethanol	(Pd(P(C$_6$H$_5$)$_3$)$_2$Cl$_2$/ HCl	Ethyl-α-chloro-propionate	37	[406]
Phenyl-vinyl-sulfo-nate	Ethanol	(P(C$_6$H$_5$)$_3$)$_2$PdCl$_2$ + (P(C$_6$H$_5$)$_3$) (C$_5$H$_{11}$N)PdCl$_2$	Phenyl-(carbethoxy)-ethane-sulfonate	55	[406]
Methyl-acrylate	Methanol	Co$_2$(CO)$_8$/Py/H$_2$	Dimethylsuccinate dimethyl-malonate dimethyl-γ-ketopimelate	75 3 13	[1024]

6.6. Olefins and Functional Derivatives in the Presence of Carboxylic Acids, Thiols, Amines or Hydrogen Chloride

W. Reppe and H. Kröper [483] isolated large amounts of propionic acid anhydride in the synthesis of propionic acid from ethylene, carbon monoxide and water. They proved their presumption that ethylene and carbon monoxide react with preformed propionic acid by repeating the experiments with propionic acid as starting material (1).

$$H_2C=CH_2 + CO + H_3C–CH_2–COOH \xrightarrow{\text{cat.}} (H_3C–CH_2–\underset{\underset{O}{\|}}{C})_2O \quad (1)$$

Formation of anhydride succeeds with Ni catalysts even at lower temperatures (230 to 250 °C) than the synthesis of propionic acid from ethylene. Thiolcarboxylic acid esters are obtained analogously by addition of thiols instead of carboxylic acids (2). Olefins, carbon monoxide and amines react to give saturated carboxylic acid amides (3) and acid chlorides are formed from hydrogen chloride and carbon monoxide in the presence of noble metal catalysts of the 8th group of the periodic table of the elements (4).

$$H_2C=CH_2 + CO + HS–R \xrightarrow{\text{cat.}} H_3C–CH_2–\underset{\underset{O}{\|}}{C}–SR \quad (2)$$

$$H_2C=CH_2 + CO + HNRR' \xrightarrow{\text{cat.}} H_3C–CH_2–\underset{\underset{O}{\|}}{C}–NRR' \quad (3)$$

$$H_2C=CH_2 + CO + HCl \longrightarrow H_3C–CH_2–COCl \quad (4)$$

Reactions (2) and (3) require higher temperatures (280 °C). Instead of ammonia and carbon monoxide, formamide also can be used [501]. In table 49 some derivatives of saturated carboxylic acids obtained from carboxylic acids, thiols, amines and HCl are listed. Reaction (4) proceeds at moderate temperatures e.g. at 170 °C [531] or at 100 °C [512, 523]. The catalyst can be recycled several times [512].

6.7. Alcohols, Ethers and Carboxylic Acid Esters

Instead of olefins, alcohols, ethers, esters and certain lactones are also suited as starting materials in the syntheses of carboxylic acids or esters (mechanism see page 79). Alcohols react to give saturated carboxylic acids with one more C-atom than the starting material. The reaction proceeds with Ni catalysts in the presence of halogens at 280 to 300 °C and pressures of 200 to 700 atm.

Table 49. *Reaction of olefins and carbon monoxide with carboxylic acids, thiols, amines or hydrogen chloride in the presence of transition metal carbonyls*

Olefin	Other compound	Catalyst	Products	Yield (%)	Ref.
Ethylene	Propionic acid	Ni or Co	Acetic anhydride	63	[533–536]
Ethylene	Hydrogen chloride	Ru or Rh halide	Propionyl chloride		[531]
Ethylene	Hydrogen chloride	$Pd(P(C_6H_5)_3)_2Cl_2$	Propionyl chloride		[512, 532]
Octene	Ethyl-mercaptan	$Ni(CO)_4$	Ethyl thio α-methyl-caprylate	15	[537]
Butadiene	Hydrogen chloride	$Pd(P(C_6H_5)_3)_2Cl_2$	Butene-2-carb-oxylic acid chloride		[532]
Octa-decene	Ethyl mercaptan	NiI_2	Ethyl thio α-methyl-stearate	23	[537]
Ethylene	Thiophenol	$Ni(CO)_4$	Phenyl thio propionate		[537]
Ethyl-undece-noate	Ethyl mercaptan	$Ni(CO)_4/NiI_2$	Ethyl thio carbethoxy-dodecanoate	30	[537]
Ethylene	Ammonia	$HCo(CO)_4$	Propionamide	90	[540]
Propene	Aniline	$HCo(CO)_4$	Butyric acid anilide + iso-butyric acid anilide (4:1)	87	[538]
Octene-1	Aniline	$Co_2(CO)_8$	Nonanoic acid anilide	69	[539]
Cyclo-hexene	Ammonia	Co	Cyclohexane carboxylic acid amide	90	[540]
Unde-cenoic-acid	Ammonia or form-amide	Ni	Dodecanoic acid diamide	19–70	[533]

Iron carbonyls are also active catalysts. However, more effective are cobalt catalysts [375, 541–543] and even at 180 °C a high reaction rate is observed. Addition of iodine increases the reaction rate remarkably.

Recently, modified rhodium carbonyls have been reported to be superior catalysts [1009–1012].

Secondary as well as tertiary alcohols are more reactive than primary ones. The main reaction (1) — illustrated with methanol — is accompanied by side reactions [543], among them (2, 3, 4, 5) should be mentioned.

$$H_3COH + CO \xrightarrow{\text{HCo(CO)}_4,\ I_2} H_3CCOOH \tag{1}$$

$$H_3CCOOH + CH_3OH \rightleftarrows H_3CCOOCH_3 + H_2O \tag{2}$$

$$2\,CH_3OH \rightleftarrows H_3C-O-CH_3 + H_2O \tag{3}$$

$$CO + H_2O \rightleftarrows HCOOH \tag{4}$$

$$CO + H_2O \rightleftarrows CO_2 + H_2 \tag{5}$$

The side reactions (2) and (3) can be largely suppressed by adding water to the reaction mixture. Moreover, formation of water gas according to (5) is inhibited by the presence of larger amounts of water.

According to Mizoroki *et al.* [1017] alkyl halides have a retarding influence on the reaction. Consequently their concentrations should be kept low.

Also the side reactions caused by free hydrogen, as for instance the homologization and formation of aldehyde from methanol, are avoided.

In industrial processes the reaction products resulting from (2), (3) and (4) can be recycled to the reactor because of the reversibility of the three reactions [543].

As can be seen from table 50, the reaction is not restricted to monools but can also be applied to diols. Straight chain diols with primary OH-groups yield preferably straight chain dicarboxylic acids.

Also anhydrides or esters [549] can be obtained applying individual reaction conditions (in analogy to the carbonylation of olefins).

Unsaturated alcohols of the allyl alcohol type react in the presence of $PdCl_2$ or bis-triphenylphosphine palladium dichloride to give unsaturated esters [448, 550].

$$H_2C=CH-CH_2OH + CO + HOR \xrightarrow{\text{PdCl}_2} H_2C=CH-CH_2-COOR + H_2O$$

Allyl vinylacetate is the main product in the reaction of allyl alcohol with carbon monoxide in the presence of tris-(tri-(p-fluorophenyl)-phosphine)-platinum at 200 °C without addition of another alcohol. At a reaction temperature of 250 °C allyl crotonate is obtained by thermal isomerization [551].

$$2\,H_2C=CH-CH_2-OH + CO \xrightarrow[220\,°C]{\text{cat.}}$$

$$\underset{\overset{\|}{O}}{H_2C=CH-CH_2-C}-O-CH_2-CH=CH_2$$

$$\Big\downarrow 250\,°C$$

$$\underset{\overset{\|}{O}}{H_3C-CH=CH-C}-O-CH_2-CH=CH_2$$

Table 50. *Carbonylation of alcohols*

Alcohol	Products	Yield (%)	Ref.
Methanol	Acetic acid	93	[312, 347, 370, 480, 490, 543, 544, 1009, 1010, 1012]
Ethanol	Propionic acid	82	[347, 370, 480, 545]
Propanol	n- and Isobutyric acid (4:1)	82	[347, 546]
Propanol-2	Isobutyric acid	68	[480]
Butanol	2-Methyl butyric acid	93	[222, 546, 547]
Butanol-2	2-Methyl butyric acid	70	[547]
Isobutanol	Isovaleric acid + trimethyl acetic acid	57	[347]
Pentanol	α-Methyl valeric acid	16	[547]
Neopentanol	C₆-Acids	21	[547]
2-Methylbutanol-2	β,β-Dimethyl butyric acid	35	[547]
Hexanol	2-Methylhexanoic acid	55	[547]
2-Ethyl butanol-1	β-Methyl, β-ethyl butyric acid	40	[547]
Heptanol	2-Methylheptanoic acid	33	[547]
Heptanol-2	2-Methyl heptanoic acid	70	[547]
Octanol	2-Methyl octanoic acid	30	[547]
Octanol-2	2-Methyl octanoic acid	76	[547]
3-Cyclohexyl-propanol	3-Cyclohexyl-2-methyl propionic acid	49	[547]
Cyclopentanol	Cyclopentanoic acid	84	[547]
4-Methylcyclo-hexanol	4-Methylcyclohexanoic acid	53	[547]
Decalol-2	Decahydro-1 and 2-naphthenoic acid	77	[547]
2-Phenyl-ethanol	Ethyl benzene	12	[547]
3-Phenyl-propanol	n-Propyl benzene	40	[547]
4-Phenyl-butanol	n-Butyl benzene	40	[547]
Ethane-1,2-diol	Succinic acid		[347]
Butane-1,2-diol	Adipic acid and methyl glutaric acid (2:3)	12.5	[347]
Butane-1,4-diol	Adipic acid	69	[347, 548]
Pentane-1,5-diol	Pimelic acid, α-methyl valeric acid	10	[347, 547]
Hexane-1,6-diol	Suberic acid, methyl hexanoic acid	30	[347, 547]
Decane-1,10-diol	Decane-1,10-dicarboxylic acid	57	[347]
Dodecane-1,12-diol	Dodecane-1,12-dicarboxylic acid	60	[347]
Tetradecane-1,14-diol	Tetradecane-1,14-dicarboxylic acid	66	[347]

Table 51. *Carbonylation of ethers, carboxylic acid esters and lactones*

Starting material	Catalyst	Products	Yield (%)	Ref.
Dimethylether	NiI$_2$	Methyl acetate	37	[347]
Ethylene oxide	Co	Ethyl-β-hydroxypropionate	13	[297]
Propylene oxide	Co	Methyl-β-hydroxybutyrate	40	[554]
Propylene oxide	Co	Methyl-β-hydroxybutyrate	11	[297]
Cyclohexene oxide	Co	Ethyl 2-hydroxy cyclohexane-carboxylate	23	[297]
Epichlorohydrin	Co	Ethyl γ-chloro-β-hydroxy-propionate	3	[297]
Tetrahydrofuran	Ni	Adipic acid, valeric acid δ-valerolactone	74 17 9	[348]
Tetrahydrofuran	Co	Valerolactone	75	[348]
2-Methyltetrahydrofuran	Ni	2-Methyladipic acid		[348]
2,5-Dimethyltetrahydrofuran	Ni	2,5-Dimethyl adipic acid		]348]
Tetrahydropyran	Ni	Pimelic acid		[348]
Dioxane	Ni	α-Methyl glutaric acid	20	[348]
α-Methyl-γ-butyrolactone	Ni	Adipic acid, valeric acid	65–75	[348]
Methylacetate	Ni	Acetic anhydride	43	'[355]
β-Propiolactone	Co	Succinic anhydride	29	

Analogously to alcohols, ethers, esters and lactones can be reacted. As to the mechanism it can be postulated that first cleavage of the C—O bond under formation of alkylcobalt or alkylnickel carbonyls or, in the presence of halogen acids, formation of halogenides occurs by action of the catalyst.

$$H_3C-O-CH_3 + HCo(CO)_4 \longrightarrow H_3C-Co(CO)_4 + CH_3OH \quad (1)$$

$$H_3C-O-CH_3 + HI \longrightarrow CH_3I + CH_3OH \quad (2)$$

Subsequent reactions proceed in analogy to the mechanism given on page 80 of this chapter.

It can be illustrated by the following overall equations:

$$ROR' + 2CO + H_2O \xrightarrow{cat.} RCOOH + R'COOH$$

$$ROR' + CO \xrightarrow{cat.} RCOOR'$$

$$\text{(cyclic ether)} + 2CO + H_2O \xrightarrow{cat.} HOOC-(CH_2)_4-COOH$$

$$H_2C\underset{O}{\overset{}{\diagdown}}CH_2 + CO + H_2O \xrightarrow{cat.} H_2C-CH_2-COOH \underset{OH}{} \xrightarrow{-H_2O} H_2C=CH-COOH$$

$$HCOOR + CO \xrightarrow{cat.} (RCO)_2O$$

Cyclic ethers react faster than alicyclic ones [347, 348, 553]. Highly reactive are epoxides. Their reaction products — hydroxycarboxylic acids or the corresponding esters — react to give unsaturated acids or esters by dehydration. Under mild conditions the hydroxyesters can be isolated as such. McClure reported a 53% yield of 4-chloro-3-hydroxy butyrate obtained in the carbonylation of epichlorohydrin [1023].

6.8. Saturated Aldehydes

As mentioned in the discussion on hydrogenation of aldehyde groups, under hydroformylation conditions which may result to a certain extent in formation of formiates, also carbon-oxygen double bonds are accessible to reactions with carbon monoxide (see page 67).

Under the conditions of hydrocarboxylation good yields of hydroxy carboxylic acids and their derivatives can be obtained. Thus, formaldehyde reacts with carbon monoxide and water to give glycolic acid [556].

$$HCHO + CO + H_2O \xrightarrow{\text{cat.}} \underset{\overset{|}{OH}}{CH_2-COOH}$$

Active catalysts are Ni, Co or Fe halides. The activity falls along the series Ni > Co > Fe. Among the halides iodides are most effective. Reaction temperatures applied are between 150 to 275 °C at 150 to 650 atm. Conversion is satisfactory. Thus, formaldehyde reacts with NiI_2 at 200 °C and 610 atm with 47% conversion to give 90% of glycolic acid besides 5% of formic acid and 5% of methanol [556].

I. Kato et al. [557] described the synthesis of acetic acid from aqueous formaldehyde and carbon monoxide in the presence of NiI_2 in 82% yield. In this case it cannot be decided precisely whether acetic acid is formed via glycolic acid and subsequent hydrogenation or via the prior hydrogenation of formaldehyde to methanol.

6.9. Halides

The last reaction discussed in this section will be the carbonylation of halides.

The reaction proceeds with aliphatic as well as with aromatic halogen compounds. The mechanism of this reaction is formulated by R. F. Heck [558–561] according to (1)–(4):

$$RX + Ni(CO)_4 \longrightarrow RNi(CO)_2X + 2\,CO \qquad (1)$$
$$RNi(CO)_2X + CO \longrightarrow RCONi(CO)_2X \qquad (2)$$
$$RCONi(CO)_2X + 2\,CO \longrightarrow RCOX + Ni(CO)_4 \qquad (3)$$
$$RCOX + HOR \longrightarrow RCOOR + HX \qquad (4)$$

resp.

$$RCOX + HNRRR' \longrightarrow RCONRR' + HX \qquad (4a)$$

Table 52. *Carbonylation of halides, sulfates and sulfonates*

Starting material	Products	Ref.
Methyl iodide	Methyl acetate	[121]
Methyl p-toluene-sulfonate	Methyl acetate	[121]
Diethyl sulfate	Methyl propionate	[121]
Amyl iodide	Ethyl capronate	[121]
1-Chlorooctane	Methyl pelargonate, methyl α-methylcaprylate	[121]
1-Iodooctane	Methyl pelargonate	[121]
2-Iodooctane	Methyl α-methylcaprylate	[121]
Allyl bromide	Methyl vinylacetate	[121, 562]
Allyl chloride	Methyl vinylacetate	[550]
Allyl chloride	Crotonic acid	[438]
Allyl chloride	Vinylacetyl chloride	[563]
Crotyl chloride	3-Pentenoic acid	[438]
3-Chlorobutene-1	3-Pentenoyl chloride	[563]
Crotyl chloride	3-Pentenoic acid	[438]
3-Chlorobutene-1	3-Pentenoyl chloride	[563]
1-Chloro-2-methyl-propene-2	3-Methylbut-3-enoyl chloride	[563]
Propargyl chloride	Ethyl butadi-2,3-enoic acid	[564, 565]
Propargyl chloride	Methyl itaconate 66% + methyl-3-chlorobut-3-enoic acid-1 (traces)	[566]
3-Bromocyclooctene	Ethyl cyclooctene-2-carboxylate-1	[529]
1-Chloro-4-cyanobutene-2	Dihydromuconic acid mononitrile	[438]
Benzyl bromide	Methyl phenylacetate	[121]
α,α-Dichloro-p-xylene	Dimethyl p-tolyldiacetate	[121]
α-Chloronaphthalin	Methyl α-naphthylacetate	[121]
Methyl α-chloroacetate	Dimethyl malonate	[121]
Methyl α-bromopropionate	Dimethyl methylmalonate	[121]
Iodobenzene	Methyl benzoate	[567]
Iodobenzene	Benzoic acid	[568]
Chlorobenzene	Methyl benzoate	[569–575]
p-Chlorotoluene	p-Toluic acid	[567]
m-Chloro iodobenzene	Methyl m-chlorobenzoate	[567]
p-Dichlorobenzene	p-Chlorobenzoic acid + terephthalic acid	[568–575]
p-Dichlorobenzene	Methyl p-chlorobenzoate + methyl terephthalate	[569]
m-Dichlorobenzene	Isophthalic acid	[570–575]
p-Dibromobenzene	p-Bromobenzoic acid + terephthalic acid	[577]
o-Dichlorobenzene	o-Chlorobenzoic acid + phthalic acid	[568]
p-Chloroanisol	Anisic acid	[570–575]
1-Chloronaphthalene	Naphthalene carboxylic acid-1	[568]
1-Iodonaphthalene	Methyl naphthalene carboxylate-1	[567]
Benzyl chloride	Phenylacetic acid anilide	[121]

The conversion is catalyzed by nickel as well as by cobalt, iron, palladium [438, 550, 562] and rhodium compounds [550]. Reaction temperatures in the range of 200–300 °C and pressures from 600 to 1000 atm are applied.

More moderate conditions can be used with $NaCo(CO)_4$ and equimolar amounts of tertiary amines or alcohols related to the halides. Under these conditions the reaction proceeds even at 0–100 °C under atmospheric pressure [121].

$$RX + CO + R'OH + B \xrightarrow{\quad Co(CO)_4{}^{\ominus}\quad} RCOOR' + HB^{\oplus}X^{\ominus}$$

Yields up to 80 % are obtained. Instead of halides also sulfates or sulfonates can be used. Esters are formed in the presence of alcohols or amides in the presence of amines (see table 52). Acid chlorides are available by reaction of carbon monoxide with allyl chloride in the absence of solvents [121, 562].

7. Industrial Carbonylation Operations and Economic Aspects of Their Reaction Products

Although numerous carbonyl compounds available by Reppe reactions are of high economic interest, for instance the acrylates and saturated carboxylic acids, capacities of carbonylation processes are much lower than capacities of hydroformylation processes. In many cases it may be due to the fact that the same products can be obtained by more attractive processes starting with easily available olefins.

Actual operating capacities of Reppe carbonylation processes are difficult to estimate since only a few data are available in the literature. However, it is known that some of the syntheses are carried out on an industrial scale, e.g. the synthesis of acrylates from acetylene, carbon monoxide and alcohols (BASF) [1004, 1005], the acetic acid synthesis from methanol and carbon monoxide and the synthesis of higher molecular weight saturated carboxylic acids from olefins, carbon monoxide and water. Propionic acid (30,000 tons/year) and to a smaller extent heptadecanoic dicarboxylic acid are manufactured via the carbonylation route at BASF. Butanol is made from propylene in Japan [1003, 1004].

A variation of the conventional acrylate synthesis was developed by Rohm and Haas in the USA [578, 579] and Montecatini in Italy [580]. This process tries to combine the advantages of both the pressureless stoichiometric procedure and the low $Ni(CO)_4$ catalyst concentration of the catalytic procedure.

At atmospheric pressure and temperatures of 30–50 °C acrylates are obtained in yields of 80–90 % by this method of operation. Only 15 % of the required carbon monoxide originates from $Ni(CO)_4$.

The avoidance of the dangerous working with acetylene under pressure, requiring special techniques for safe handing [581, 582], in particular lowers the cost of the end products.

Recently the economic aspects of the acrylate synthesis from acetylene, carbon monoxide and alcohols were reviewed in European Chemical News [583]. The technology of the acrylate synthesis was reviewed in detail by M. Sittig [584].

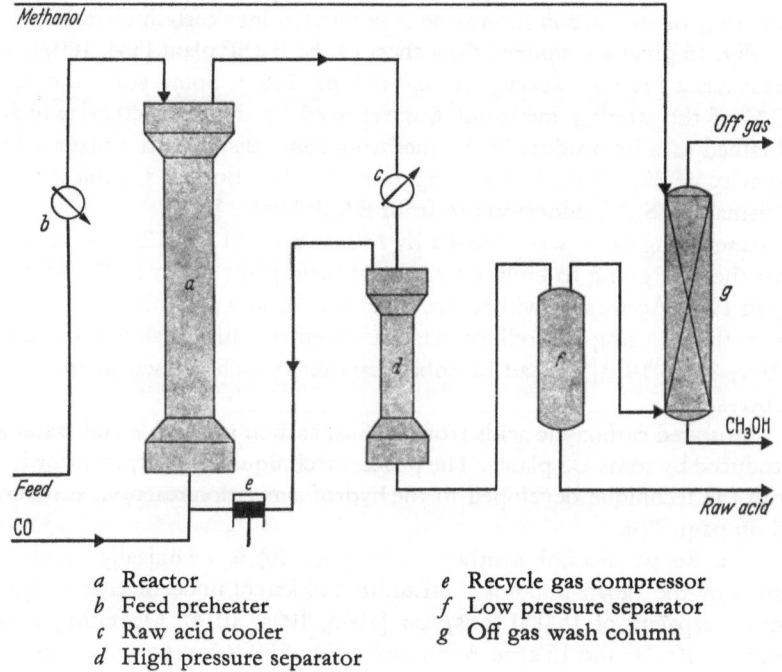

a Reactor
b Feed preheater
c Raw acid cooler
d High pressure separator

e Recycle gas compressor
f Low pressure separator
g Off gas wash column

Fig. 15. Flow sheet of BASF acetic acid plant

The acrylate capacity of the western world was estimated to be about 165,000 tons in 1966, about 110,000 tons of which were made through the Reppe route.

The acetic acid process was developed at BASF in Ludwigshafen to plant-scale application [543, 1007, 1008, 1012]. It operates with cobalt/iodine as catalyst and with addition of water at about 250 °C and 750 atm in the carbonylation reactor [1012]. The greatest difficulty in the development was presented by the corrosiveness of the reaction mixture because special steels as well as platinum, titanium and tantalum linings corrode. The problem was solved by the use of Hastelloy C (Ni, Mo, Cr) [585, 1007, 1008, 1012] and the somewhat less stable Hastelloy B (Ni, Mo, Cr) [585].

Conversion proceeds in the usual cylindrical high pressure reactors with use of recirculating tubes. The energy for recirculating is provided by the gas. The heat of reaction liberated is 530,000 kcal/ton of acetic acid and covers nearly the whole heat demand of the plant.

The degree of catalyst recovery is very high. The cobalt is regenerated almost 100 per cent. Yield of acetic acid is 90% expressed in terms of methanol and 70% based on carbon monoxide. 3.5% of the starting methanol forms methane and 4.5% forms liquid by-products. 2% is lost in off-gas, 10% of the carbon monoxide is converted into carbon dioxide.

Fig. 15 gives a simplified flow sheet of the BASF plant [534, 1007]. The plant has a present capacity of 40,000 tons [1006]. Some years ago up to 70% of the starting methanol was replaced by dimethyl ether, which is obtained as a by-product in the methanol synthesis. Another plant with a capacity of 30,000 tons/year is operated by The Borden Chemical Co. in Geismar (U.S.A.) under license from BASF [1007, 1008].

Interesting news was released by Monsanto [1011, 1012] who reported that they are going to build a large acetic acid plant at Texas City for start-up in 1970. Acetic acid will be manufactured by low pressure carbonylation of methanol using a rhodium catalyst together with a halogen promotor [1009, 1010, 1013] instead of cobalt catalyst, which is used in the BASF process.

Saturated carboxylic acids from olefins, carbon monoxide and water are produced by some US plants. The process technique is closely in accordance with the technique developed in the hydroformylation reaction, as described on page 70ff.

The Reppe alcohol synthesis (see page 40) is technically applied in Japan by the Japan Butanol Co. Ltd. in Yokkaichi under license of BASF with a capacity of 15,000 tons/year [1003, 1004, 1014]. Operating conditions are 100 °C and 15 atm. A butanol yield of 90% is reported which consists of 85% of n- and 15% of iso-butanol.

III

Carbonylation with Acid Catalysts
Koch Reaction

1. General Remarks

The synthesis of carboxylic acids from olefins, carbon monoxide and water catalyzed by carbonyls was discussed in detail in the previous chapter. The same components also react in the presence of acid catalysts to give carboxylic acids. This process was patented by DuPont in the thirties.

If the starting materials -olefin, water and acid catalyst, e.g. H_2SO_4, H_3PO_4, HCl or $ZnCl_2$- are reacted simultaneously, reaction occurs at temperatures between $100-350\ °C$ and pressures from 500 upward to 1000 atm [725–731]. Analogous reactions are described in many patents [375, 593].

These drastic reaction conditions were a strong disincentive to industrial application of the reaction because of the corrosive catalysts used. Also other catalysts such as BF_3 could not reduce the severity of the reaction conditions [586, 587]. Yields obtained even under such severe reaction conditions were too low for industrial application with the single exception of tetramethylethylene which reacted in the presence of $BF_3 \cdot 3\ H_2O$ at 600 atm and 75 °C in nearly quantitative yield to give a carboxylic acid, as described by Ford et al. [586].

The decisive success in these syntheses was made by H. Koch, W. Gilfert and W. Huisken [588, 589] who found that high olefin conversion could be achieved at mild reaction conditions if the reaction is carried out in two stages. In the first stage the olefin reacts with the acid catalyst and carbon monoxide in the absence of water and in the second stage the complex formed from olefin, carbon monoxide and acid catalyst is hydrolysed.

$$H_2C{=}C{-}CH_3 \ \underset{2.\ H_2O}{\overset{1.\ H_2SO_4,\ CO}{\xrightarrow{\hspace{1.5cm}}}} \ H_3C{-}\overset{\displaystyle CH_3}{\underset{\displaystyle CH_3}{C}}{-}COOH$$

Under these conditions nearly all olefins and a great number of dienes, unsaturated carboxylic acid esters, unsaturated and saturated alcohols and diols, reactive cycloparaffins, saturated and unsaturated halogenated compounds, N-tertiary alkyl acyl amines, carboxylic acid esters and saturated aldehydes react at temperatures between -20 and $+80\,°C$ and carbon monoxide pressures from $1-100$ atm in yields of $85-95\%$ to carboxylic acids which contain one C-atom more than the starting material.

Starting from olefin/isoparaffin mixtures carbonylation of the isoparaffin also occurs by hydride transfer.

At atmospheric pressure and temperatures between $0-40\,°C$, Koch syntheses succeed if formic acid is used as the CO source. In this case formic acid is added simultaneously with the olefin to H_2SO_4 and dehydrated to carbon monoxide by the sulfuric acid.

Carboxylic acid esters are obtained instead of carboxylic acids if the intermediate complex formed from the starting material, carbon monoxide and catalyst is decomposed with alcohols instead of water in the second stage. Since fundamentally no other results are obtained with alcohols compared to water in view of yields or the isomer distribution, both reactions will be discussed together in the individual sections of this chapter.

2. Reaction Mechanism

H. Koch [593] formulated the mechanism of the carboxylic acid synthesis found by him via initial formation of a carbonium ion from the starting material and the acid catalyst with subsequent addition of carbon monoxide to the carbonium ion to give an acylium cation in the first reaction step (1) followed by the reaction of the acylium cation with water or alcohol in the second step (2).

$$H_2C=CHR + H^\oplus \longrightarrow H_3C-\overset{\oplus}{C}HR \xrightarrow{\ CO\ } H_3C-CHR-CO^\oplus \quad (1)$$

$$H_3C-CHR-CO^\oplus + HOR' \longrightarrow H_3C-CHR-COOR' + H^\oplus \quad (2)$$

$$R'=H,\ Alkyl$$

Eidus *et al.* [594] have suggested that in H_2SO_4 catalyzed reactions the first stage of the Koch reaction consists of the formation of an acylsulfuric acid (3) which in the second stage reacts with water or alcohols to form carboxylic acids or esters (4).

$$H_2C=CHR + H_2SO_4 + CO \longrightarrow H_3C-CHRCOHSO_4 \quad (3)$$

$$H_3C-CHRCOHSO_4 + HOR' \longrightarrow H_3C-CHR-COOR' + H_2SO_4 \quad (4)$$

$$R'=H,\ alkyl$$

However, the existence of such mixed anhydrides has not yet been experimentally proved [595].

To a large extent the Koch reaction is accompanied by double bond isomerizations and isomerization of the carbon skeleton with migration of alkyl groups, which is thought to occur through the carbonium ions (see e.g. scheme page 125).

$$H_3C-CH_2-CH_2-CH_2-CH_2-CH=CH_2 \longrightarrow$$

$$H_3C-CH_2-CH_2-CH_2-\overset{\oplus}{C}H-CH_2-CH_3 \quad \xleftarrow{\sim H}$$

$$\xrightarrow[-H^{\oplus}]{CO/H_2O} \quad H_3C-CH_2-CH_2-CH_2-CH-COOH \;|\; C_2H_5$$

$$H_3C-CH_2-CH_2-CH_2-\overset{\oplus}{C}-CH_3 \;|\; CH_3 \quad \xleftarrow{\sim CH_3}$$

$$\xrightarrow[-H^{\oplus}]{CO/H_2O} \quad H_3C-CH_2-CH_2-CH_2-CH-COOH \;|\; CH_3$$

$$H_3C-CH_2-CH_2-\overset{\oplus}{C}-CH_2-CH_3 \;|\; CH_3 \quad \xleftarrow{\sim CH_3}$$

$$\xrightarrow[-H^{\oplus}]{CO/H_2O} \quad CH_3-\overset{CH_3}{\underset{CH_3}{\overset{|}{\underset{|}{C}}}}-COOH$$

$$\xrightarrow[-H^{\oplus}]{CO/H_2O} \quad H_3C-\overset{C_2H_5}{\underset{C_2H_5}{\overset{|}{\underset{|}{C}}}}-COOH$$

The alteration of ring size which may occur in case of cyclic compounds will be discussed later.

The addition of the carboxyl group strictly follows the Markovnikov rule and no straight chain carboxylic acids are formed by addition of the carboxyl group to a terminal C-atom, except for the case where ethylene is used as feedstock and where no other than terminal addition is possible. Therefore the number of possible isomers is given under normal reaction conditions by $Z = N\text{-}2$, N being the number of C-atoms in the starting material. In practice the number of carboxylic acids formed often is higher, since, in many cases, especially at low CO-pressure, carbonylation occurs only after di- or oligomerization of the starting material and therefore larger amounts of higher molecular weight acids are obtained.

On the other hand the mineral acid also catalyzes cracking of the starting material. Thus, tertiary butyl groups are split off easily, resulting in the formation of carboxylic acids with a lower number of C-atoms than the starting material.

3. Catalysts

Suitable catalysts for Koch-syntheses are: conc. sulfuric acid, conc. phosphoric acid, HF and mixtures of BF_3 with H_2O, CH_3OH, HF, H_2SO_4 or H_3PO_4 and HF/SbF_5 mixtures [1002]. If BF_3/CH_3OH is used as catalyst, the main reaction products are the methyl esters of the carboxylic acids besides small amounts of the corresponding free acids [596].

As mentioned above the catalyst used must be free of water to obtain high yields. Best results are obtained e.g. with propylene using a 96% H_2SO_4 (H_2SO_4 usual in trade). Acid concentrations of 90% and lower decrease the yield remarkably. This is in line with results obtained with other catalysts.

Best yields with olefins from the isobutene and tetramethylethylene type are obtained with a 82–88% sulfuric acid as catalyst.

Recently Pawlenko reported the oxonium tetrafluoroborates of the type $(ROH_2)(BF_4)$ to be superior catalysts [999, 1000] allowing very elegant technical operations even with olefins which are normally difficult to carboxylate (such as ethylene and propylene) (see also the chapter on industrial applications page 144).

In the presence of larger amounts of water, formation of alcohols may occur by hydration of the olefin instead of carbonylation. In turn these alcohols can form esters with the acid reaction products [598].

As K. E. Möller showed, the catalyst has a marked effect upon the isomer distribution of the carboxylic acids produced [599].

With BF_3-containing catalysts, which cause less isomerization compared to H_2SO_4, more secondary than tertiary carboxylic acids are obtained Also H_2SO_4 yields compared to BF_3-catalysts more secondary carboxylic acids with a more centrally located carboxyl group. Moreover, the ratio of acid catalyst to olefin has a considerable influence on yield. Eidus *et al.* have found in their experiments to synthesize esters that yield increases with the amount of excess catalyst [601]. With less than the stoichiometric amount of catalyst, yields of carboxylic acids are poor; the main reaction products are alcohols formed from olefins [602].

In the reaction of branched olefins, contrary to straight chain olefins, the amount of acid (mole to mole) can be lowered and the concentration of water in the acid catalyst may be higher [603–605]. If long residence times of carbon monoxide, conc. acid and starting material are applied before the reaction is stopped by addition of water or alcohol, considerable amounts of ketones may be formed, especially starting from mixtures of olefins and paraffins [606].

$$\begin{array}{ccc} R & R'' \\ \mid & \mid \\ H_3C-C-R' + H_3C-C-R''' \\ \mid & \mid \\ {}^{\oplus}C=O & H \end{array} \longrightarrow \begin{array}{c} R \quad R'' \\ \mid \quad \mid \\ H_3C-C-C-C-CH_3 + H^{\oplus} \\ \mid \; \parallel \; \mid \\ R' \; O \; R''' \end{array}$$

Industrial applications of the described catalysts in continous processes require their good separation from the reaction product after addition of water or alcohol to the carbonyl group. This separation can be effected in the case of conc. H_2SO_4 by dilution of the homogenous reaction mixture with water. After dilution the sulfuric acid has a concentration of 60–70 % and cannot be recycled.

In the reaction of branched olefins a mixture of H_2SO_4/BF_3 can be used as catalyst which allows a nearly quantitative separation of the catalyst from the reaction product without dilution [60, 609]. In most cases the catalyst system H_3PO_4/BF_3, available by introduction of BF_3 into a 85 % H_3PO_4 up to a mole ratio of 1.2 : 0.8 to 0.8 : 1.2 [607, 608], is also suitable, and allows an excellent separation from the reaction products. Therefore this catalyst mixture is favored in industrial processes [610].

A detailed comparison of the different catalyst types is given by K. E. Möller [611].

4. Influence of Temperature and Pressure

Temperature and pressure, and hence the carbon monoxide concentration in the reaction mixture, have a remarkable influence on yield and distribution of the reaction products in Koch syntheses.

Reaction temperatures may be varied between – 20 to + 80 °C, depending on the olefin and the desired reaction product. As mentioned above, in the first stage the reaction initiates through carbonium ions formed by proton addition of the starting material. It is well known that such carbonium ions isomerize quickly and without a large energy demand into ions of higher stability [612]. Depending on the reaction temperature, this equilibrium may be achieved or the initial carbonium ions may react directly with carbon monoxide with the formation of acylium cations. Generally, low reaction temperatures yield more secondary, high reaction temperatures more tertiary carboxylic acids [613, 614].

Table 53. *Temperature dependence of the ratio of tertiary to secondary carboxylic acids; CO-pressure 100 atm*

Olefin	H_2SO_4 (96%)				$BF_3 \cdot CH_3OH$			
	tert. acid (%)		sec. acid (%)		tert. acid (%)		sec. acid (%)	
	−5°C	+15°C	−5°C	+15°C	−5°C	+15°C	−5°C	+15°C
Hexene-1	43	58	57	42	22	35	78	65
Heptene-1	54	68	46	32	26	42	74	58
Octene-1	50	64	50	36	30	43	70	57
Nonene-1					34	44	66	56
Decene-1					33	50	67	50

The influence of CO pressure is illustrated in fig. 16.

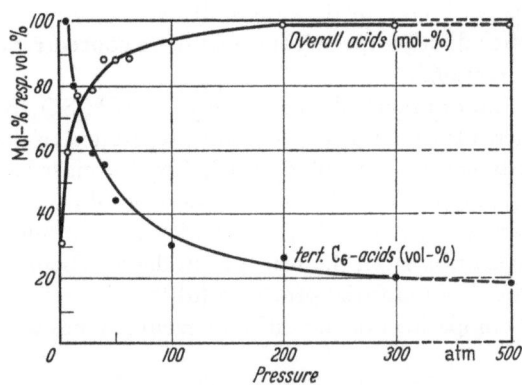

Fig. 16 [615]. The dependence of methyl group migration on CO pressure

In general it was found that an increase in pressure leads to an increase in conversion of olefin and yield of carboxylic acids [615, 616]. At atmospheric pressure and slightly above, secondary acids are the main reaction products. The strong influence of pressure, temperature and catalyst on product distribution can be seen from table 54 [600], published by K. E. Möller.

Table 54. *Influence of reaction conditions on yield and isomer distribution in cyclohexene carbonylation*

Catalyst	Temp. (°C)	CO pressure (atm)	Yield (%)* C_7-acids	Isomer distribution (%)	
				(cyclohexyl)COOH	C(cyclopentyl)COOH
HCOOH/H_2SO_4**	10	—	75	—	100
H_2SO_4	15	150	78	78	22
H_2SO_4	−13	150	72	93	7
$BF_3 \cdot CH_3OH$	15	150	80	93	7

  * Yield on cyclohexene feed
 ** Under the same conditions the isomer distribution can be altered if the stirring velocity is diminished

Generally it can be said that low temperature and high pressure favor the formation of carboxylic acids containing one C-atom more than the starting material [600, 601, 614–616]. At lower pressure and higher reaction temperatures the amount of high molecular weight carboxylic acids is increased by previous di- and oligomerization; e. g. in syntheses of esters, only polymer products could be isolated when temperatures above 75 °C were applied [601].

The influence of pressure can be seen from table 55 [618].

Table 55. *Influence of CO pressure on reaction of pentene-2 (Catalyst: 96% H_2SO_4)*

CO pressure (atm)	Yield of carboxylic acids (mole-%)	Composition of carboxylic acids according to the number of C-atoms					Composition of C_6-acids (mole; %)	
		C_6	C_{11}	C_{10}	C_{21}	C_{26}	tert.	sec.
1	31.5	11.1	25.6	19.8	18.7	24.5	100	—
5	59.0	24.3	24.7	18.2	32.8	—	100	—
10	77.5	95.0	5.0	—	—	—	79.7	20.3
20	78.0	100	—	—	—	—	64.5	35.5
30	88.0	100	—	—	—	—	60.9	40.0

At pressures higher than 20 atm. C_6-acids are formed exclusively from pentene-2. At lower pressures more higher molecular weight carboxylic acids are obtained; at a pressure of 1 atm. their amount comes to 90%. However, under special conditions larger amounts of secondary acids can be produced if the so-called formic acid method is applied [619], namely if care is taken to sustain a high CO concentration in the reaction mixture. Haaf stated that less effective stirring of the reaction mixture can result in an supersaturation of carbon monoxide in the range of 10^2. This increase is sufficient to convert large amounts of the initial secondary carbonium ions to secondary carboxylic acids before rearrangement to tertiary carbonium ions can occur (see table 56).

Table 56. *Influence of stirring velocity on yield of carboxylic acids in the Koch synthesis using formic acid as CO source*

Starting material	% Yield (first figure obtained with moderate stirring 20 cpm; figures in brackets with vigorous stirring)
Pentanol-2	30 (79) 2-Methyl pentanoic acid-2
	26 (1) Hexanoic acid-3
	26 (1) Hexanoic acid-2
Heptanol-2	17 (29) 2-Methylheptanoic acid-2
	28 (48) 3-Methylheptanoic acid-3
	9 (3) Octanoic acid-4
	23 (1) Octanoic acid-3
	8 (1) Octanoic acid-2
Cyclohexanol	14 (61) 1-Methyl cyclopentane carboxylic acid-1
	75 (8) Cyclohexane carboxylic acid

5. Solvents and Dilution Media

The Koch carboxylic acid synthesis proceeds with high reaction velocity if care is taken to stir the reaction mixture sufficiently. After formation of the acylium cations, the carboxylic acids can be separated nearly quantitatively by adding the stoichiometric amount of water if suitable catalysts are applied. Therefore in most cases solvents or dilution mediums are not required.

The number of solvents is restricted by the reactive catalysts applied. Suitable dilution mediums are so far compounds which are not, or only to a limited extent, able to form carbonium ions, e. g. CCl_4 [619], unbranched paraffins [615, 620, 621] or — in special cases and only with defined catalysts — mixtures of the above mentioned paraffins and aromatic hydrocarbons.

6. Carbonylation of Special Compounds

6.1. Olefins and Dienes

As mentioned before, nearly all olefins participate in the Koch synthesis [1002] and only olefins which are not able to undergo isomerization, such as ethylene and propylene, form consistent reaction products — propionic acid and isobutyric acid.

As a consequence of the described double bond and skeletal isomerizations, mixtures of isomeric carboxylic acids are obtained starting with straight chain and isomerizable olefins (see formula scheme page 125). The position of the double bond in the olefinic starting material is of no consequence for the composition of the isomeric products. Thus, starting from hexene-2 and hexene-3 respectively the same isomer distribution of carboxylic acids is obtained as with hexene-1 as starting material [613]. However, only carboxylic acids are formed which result in an addition of the carboxyl group following the Markovnikov rule.

Secondary and tertiary carboxylic acids are formed from unbranched olefins, whereas branched olefins react to give exclusively tertiary acids [611].

Starting from symmetrical and nonisomerizable olefins, mixtures of carboxylic acids may be obtained by di- or oligomerization or by olefin cracking prior to carbonylation under special reaction conditions (see page 126). Especially branched olefins have a tendency to dimerize and therefore it is not surprising that larger amounts of 2,2,4,4-tetramethylpentanoic acid-1 are obtained besides the expected pivalic acid starting from isobutene.

$$
\begin{array}{c}
H_3C-C=CH_2 \\
\quad\ \ | \\
\quad\ \ CH_3
\end{array}
\left[
\begin{array}{l}
\rightarrow\ (CH_3)_3-C-COOH \\[1em]
\rightarrow\ (CH_3)_3-C-CH_2-C(CH_3)_2-COOH
\end{array}
\right.
$$

Moreover, with unbranched low molecular weight olefins a sulfuric acid catalyzed disproportionation is observed. Thus, carbonylation of 2-methylbutene-1 led to formation of equivalent amounts of pivalic acid and 2-methylpentane carboxylic acid-2 besides 2-methylbutane carboxylic acid-2 and a mixture of tertiary C_{11}-acids [617]. The formation of these two acids was explained by cleavage of isodecene which was formed by dimerization of 2-methylbutene.

$$
\text{2-methylbutene-1}
\left[
\begin{array}{l}
\rightarrow\ H_3C-CH_2-C(CH_3)_2-COOH \\[1em]
\rightarrow\ \text{isodecene}
\left[
\begin{array}{l}
\rightarrow\ \text{tertiary } C_{11}\text{-acids} \\
\rightarrow\ (CH_3)_3-C-COOH \\
\rightarrow\ H_3C-CH_2-CH_2-C(CH_3)_2-COOH
\end{array}
\right.
\end{array}
\right.
$$

The rule "$Z = N - 2$", mentioned in the chapter on the reaction mechanism, for the estimation of the possible isomers starting from unbranched olefins cannot be applied in the case of cyclic monoolefins.

With cycloolefins the isomer distribution is as much dependent on reaction conditions as on ring size and ring strain [623]. The ring size of the starting material has a strong influence on rearrangement of the ring skeleton. Thus, e.g. even under optimum reaction conditions cyclooctene does not react to give the expected cyclooctane carboxylic acid. Always only 1-methylcycloheptane carboxylic acid-1 and 1-ethylcyclohexane carboxylic acid-1 [611, 624, 993] are obtained, whereas cyclopentene reacts

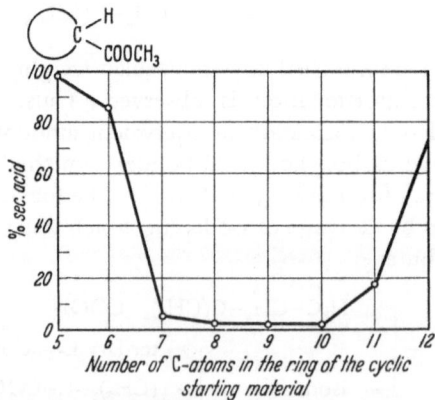

smoothly to give cyclopentane carboxylic acid [600]. Formation of dimeric acids is observed as a side reaction in the cycloolefin carbonylation but no disproportionation could be noted.

A number of rules have been derived for the carbonylation of cyclo-olefins according to the so-called formic acid method [592] (deviations from the rules relative to isomer distribution resulting from supersaturation of the reaction mixture are pointed out in chapter 4, page 130).

1 st) Tertiary acids are formed nearly exclusively, except from such cyclo-olefins that are unable to form these acids because of their ring strain.

2nd) Branched cyclohexane ring systems are formed if possible.

The ratio of secondary to tertiary acids obtained from cyclic olefins with different ring sizes can be seen from fig. 17.

Fig. 17 [600]. Methylesters from cycloalkenes

Cyclopentene reacts exclusively to give cyclopentane carboxylic acid. Formation of methyl-substituted four-membered rings has not been observed. Also cyclohexene reacts to give nearly 90 % of cyclohexane carboxylic acid, whereas cycloolefins containing 8, 9 and 10 C-atoms in the ring generally do not form secondary acids. The performance of the methyl-substituted cycloalkenes can be seen from fig. 18.

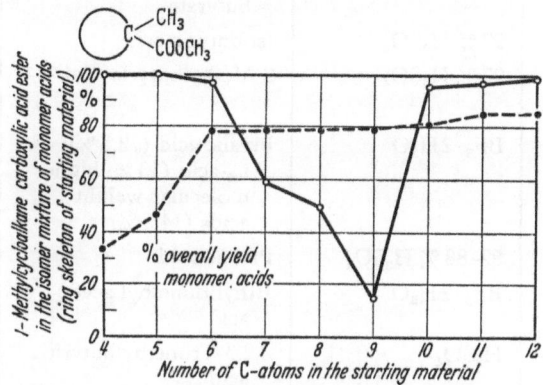

Fig. 18 [600]. Methylesters from 1-methylcycloalkenes
(Catalyst $BF_3 \cdot CH_3OH$, 150 atm CO, 15 °C)

Branching at the ring skeleton favors formation of carboxylic acids retaining the original ring size. Cycloolefins containing 4, 5, 6, 10, 11 and 12 C-atoms in the ring react exclusively with retention of the ring skeleton. In the range of C_7 to C_9-cycloolefins, which are able to isomerize, the nine-membered ring has the highest stability. In case of four- and five-membered rings dimerization of the starting material is the preferred reaction.

Diolefins with widely separated substituted double bonds react to give dicarboxylic acids in low yields [1001]. Thus, from 2,11-dimethyldodeca-diene-1,11 2,11-dimethyldodecane dicarboxylic acid is obtained in approximately 30 % yield [620]. Under the conditions of Koch syntheses conjugated dienes polymerize in the presence of the strong acid catalyst [1001]. Therefore, conjugated dienes contained in mixtures of unsaturated compounds should be separated before carbonylation. This can easily be effected by selective hydrogenation to the corresponding olefins [625].

Highly unsaturated cyclic compounds likewise do not react to give di- or tricarboxylic acids, but tertiary monocarboxylic acids are obtained with CO/H_2O by a transannular reaction with hydride transfer and formation of bi- or tricyclic systems at the bridgehead. Thus, cyclodecadiene-1,5 reacts to give a mixture of cis- and trans decalin carboxylic acid [624] and cyclododecatriene-1,5,9 reacts to give a mixture of isomeric tertiary per-hydroacenaphthene carboxylic acids.

Table 57. *Hydrocarboxylation of olefins and dienes via the Koch route*

Starting material	Catalyst	Reaction products	Yield (%)	Ref.
Ethylene	$HF + BF_3$ $(H_3O)(BF_4)$	Propionic acid Ethyl α-methyl-α-ethyl-butyrate	30 ~30	[627] [1000]
Propene	97% H_2SO_4	Isobutyric acid	90	[602]
Butene-1	97% H_2SO_4	2-Methylbutyric acid	90–95	[602]
Butene-2				
Isobutene	$BF_3 \cdot 2H_2O$	Pivalic acid (72.5%) C_9-acid (13%) higher molecular weight acids (14.5%)		[628]
	82–88% H_2SO_4	Pivalic acid		[629]
2-Methyl-butene-1	$BF_3 \cdot 2H_2O$	Ethyl dimethyl acetic acid	75	[628]
2,3-Dimethyl-butene-1	H_2SO_4	2,2,3-Trimethylbutyric acid	70	[602]
2-Ethyl-butene-1		2-Methyl-2-ethyl-butyric acid		[602]
Pentene-1	97.5% H_2SO_4	2,2-Dimethylbutyric acid (97%) 2-Ethylbutyric acid 2-Methylvaleric acid		[184]
Pentene-2	97% H_2SO_4	2,2-Dimethylbutyric acid, 2-Methylvaleric acid, Methylbutyric acid		[995]
2-Methyl-pentene-1	$H_2SO_4/BF_3/H_3PO_4$	2,2-Dimethylvaleric acid C_{13}-acids	80	[602]
3-Methyl-pentene-2		2-Methyl-2-ethyl-butyric acid		[602]
2-Ethyl-pentene-1	$BF_3 \cdot 2H_2O$	C_8-acids	62	[628]
Hexene-1	H_3PO_4	C_7-acids (87%) C_{10}-acids (13%)		[630]
Diisobutylene	82–88% H_2SO_4	Isononanoic acid		[629]
Propylene tetramer	$(H_3O)(BF_4)$	C_{13}-acid	96	[999]
Cyclopentene*	H_2SO_4	Cyclopentane carboxylic acid cis-Decalin carboxylic acid (9 : 1)	6	[591]

Table 57 (continued)

Starting material	Catalyst	Reaction products	Yield (%)	Ref.
Cyclohexene*	H_2SO_4	1-Methylcyclopentane-carboxylic acid	75 81–85	[591] [994]
Cyclooctene*	H_2SO_4	1-Ethylcyclohexane carboxylic acid higher molecular weight acids (36 : 64)	45	[991]
2,5-Dimethyl-hexa-diene-1,5	H_2SO_4	2,2,5,5-Tetramethyl-adipic acid Mono carboxylic acids	2–3 20–30	[620]
Dihydro-dicyclo-pentadiene	H_2SO_4	Exo-tricyclo-(5.2.1.0)-decene-2-carboxylic acid	74	[632]
2,11-Dime-dodecadiene		2,11-Dimethyldodecane-2,11-dicarboxylic acid	30	[620, 631]

* Formic acid method

6.2. Unsaturated Carboxylic Acids

Whereas starting from diolefins dicarboxylic acids are obtained only in small yields, unsaturated acids such as undecylenic acid and oleic acid react to give dicarboxylic acids in higher yields [591, 593, 620, 633, 998].

As was to be expected from experiments with olefins, unsaturated acids with a higher molecular weight also yield isomer mixtures of dicarboxylic acids. Schauerte [620] proved the reaction product of undecylenic acid to be an isomer mixture of seven C_{12}-dicarboxylic acids. The ratio of primary-secondary to primary-tertiary carboxylic acids was about 1:1. Under the conditions of the formic acid method only primary-tertiary acids are formed.

Recently Weintraub et al. [635] reported on the very interesting synthesis of succinic acid from acrylic acid:

$$H_2C=CH-COOH + CO \xrightarrow[\text{2. } H_2O]{\text{1. } SO_3/H_2SO_4} HOOC-CH_2-CH_2-COOH$$

The reaction failed completely under the usual reaction conditions with sulfuric acid as catalyst, but realizes a 85% yield with oleum instead of H_2SO_4 at 45 °C and a CO pressure of 85 atm.

Table 58. *Isomer distribution of the reaction products in the carboxylation reaction of undecylenic acid*

Reaction under pressure (catalyst conc. H_2SO_4 (96%), temp. 3–16 °C, 180–320 atm., CO, heptane as dilution medium)		Formic acid method (catalyst conc. H_2SO_4 (96%), temp. 13–16 °C)	
α-Methylundecane diacid	9 %		
α-Ethylsebacic acid	25.4%		
α-n-Propylazelaic acid	14.8%		
α,α-Dimethylsebacic acid	20.2%	α,α-Dimethylsebacic acid	47%
α-Methyl-α-ethylazelaic acid	10.0%	α-Methyl-α-ethylazelaic acid	24%
α-Methyl-α-n-propyl suberic acid	8.2%	α-Methyl-α-n-propyl-suberic acid	23%
α-Methyl-α-n-butylpimelic acid	3.4%	α-Methyl-α-n-butylpimelic acid	6%

6.3. Paraffins

Paraffins are also suitable starting materials in the manufacture of carboxylic acids or esters via the Koch syntheses route, assuming the formation of carbonium ions under the reaction conditions applied. It succeeds easily, especially with the reactive cyclopropanes [592, 636] and cyclobutanes [636]. Under the action of H_2SO_4 they are converted into non-cyclic carbonium ions which then react in the usual manner to give saturated carboxylic acids. The reactivity of the small strained rings falls along the series cyclopropane > alkyl substituted cyclopropane > cyclobutane > alkyl substituted cyclobutane [636]. Cyclopentane does not react under the action of sulfuric acid but cyclopentane as well as cyclohexane react in the presence of HF/SbF_5 [635], which, compared to H_2SO_4, is a more effective agent to form carbonium ions from paraffins.

In table 59 the results published so far on carboxylation of cyclopropanes and cyclobutanes can be seen.

Also in non-cyclic, branched hydrocarbons cleavage of the carbon chain may occur under formation of carbonium ions, e. g. by depolymerization and disproportionation, as discussed in the previous chapter. Thus, under suitable reaction conditions diisobutylene may yield 2 moles of pivalic acid. Aside from cracking of C–C bonds, carbonium ions may also be obtained from paraffins by hydride transfer.

Low molecular isoparaffins in sulfuric acid are especially suited as hydride ion donors. Under the conditions of the Koch syntheses tertiary carboxylic acids are formed [598, 606, 638–640].

Table 59. *Hydrocarboxylation of reactive cycloaliphatic compounds in the presence of sulfuric acid*

Starting material	Temp. (°C)	Reaction product	Yield (%)	Ref.
Cyclopropane	20	Isobutyric acid	80	[636]
Methylcyclopropane	30	2-Methylbutyric acid	78	[592, 636]
1,1-Dimethylcyclo-propane	50	2,2-Dimethylbutyric acid	60	[636]
n-Propylcyclopropane	20	2,2-Dimethylvaleric acid and 2-methyl-2-ethylbutyric acid	82	[184, 592]
Cyclobutane	50	2-Methylbutyric acid	30	[636]
Methylcyclobutane	70	2-Methylvaleric acid	17	[636]

The isoparaffin is reacted with a carbonium ion as hydride ion acceptor. In this reaction a tertiary H-atom of the isoparaffin is transferred to the prior formed carbonium ion. The new carbonium ion obtained by transfer from the isoparaffin reacts with CO to give carboxylic acids [640].

$$RH + R_1^\oplus \longrightarrow R^\oplus + R_1H$$
$$R^\oplus + CO + H_2O \longrightarrow RCOOH$$

Thus, in this reaction isoparaffins may be converted into carboxylic acids in about 75 % yield.

Starting from adamantane and tert. butanol as hydride ion acceptor adamantane carboxylic acid is formed even in 80 % yield. An 85 % yield of adamantane dicarboxylic acid is reported in a patent of DuPont [793].

The tendency to undergo the hydride transfer reaction (and therefore the yield of carboxylic acids) falls as we pass from lower to higher homologs of isoparaffins. Of course, yield largely depends on the structure of the hydride ion acceptor. The following may be used as hydride ion acceptors: olefins, alcohols or alkyl chlorides. Most efficient are tert.- and 2-butanol [640] as well as isobutanol [641] and triisobutylene [641]. Apart from H_2SO_4, HF may be used as catalyst in the carboxylation of isoparaffins. Yields however, are lower [638, 639]. Carboxylic acids with a tertiary hydrogen atom react in the same manner as dicarboxylic acids.

6.4. Alcohols and Diols

Primary as well as secondary and tertiary alcohols are able to form carbonium ions in the presence of mineral acid catalysts. Thus, they can be used as starting materials in the Koch carboxylic acid syntheses [591, 601, 643–649]. These compounds can be reacted under pressure as well as under the conditions of the formic acid method. With but few exceptions the

Table 60 [640]. *Hydrocarboxylation of isoparaffins by hydride transfer* (Catalyst H_2SO_4)

Starting material	Olefins or alcoholes (hydride ion acceptor)	Temp. (°C)	Yield of carboxylic acids (%) from	
			Isoparaffin	Olefin or alcohol
Iso-pentane	Cyclohexene	15–25*	37% 2,2-Dimethylbutyric acid	52% Cycl. C_7-acids (90% methylcyclopentane carboxylic acid-(1))
2-Methylpentane	Cyclohexene	15–25	21% C_7-acids (no cycl. acids)	59% C_7-Acids (cyclic)
2,3-Dimethylbutane	Methylcyclohexene	15–25	22% 2,2,3-Trimethylbutyric acid	33% 1-Methylcyclohexane carboxylic acid-1
2,3-Dimethylbutane	Heptene-1	15–25	31% 2,2,3-Trimethylbutyric acid	46% C_8-Acids
Methylcyclopentane	Propene	15–25	28% 1-Methylcyclopentane carboxylic acid-1	1% Isobutyric acid
n-Hexane	Cyclohexene	15–25		79% Cycl. C_7-acids (no aliph. acids)
Methylcyclpentane	Iso-propanol	15–25	18% 1-Methylcyclopentane carboxylic acid-1	1% Isobutyric acid
Cyclohexane	Tert.-butanol	15–25	7% C_6- and C_7-acids, no aliph. C_7-acids	54% Pivalic acid
Cyclohexane	Butanol-2	15–25	16% 1-Methylcyclopentane carboxylic acid-1 3% cyclohexane carboxylic acid	13% 2-Methylbutyric acid
trans Decalin	Butanol-2	15–25	8% Decalin carboxylic acid-9 (mainly cis)	12% 2-Methylbutyric acid

Table 60 (continued)

Starting material	Olefins or alcohol (hydride ion acceptor)	Temp. (°C)	Yield of carboxylic acids (%) from	
			Isoparaffin	Olefin or alcohol
trans Decalin	tert. Butanol	15–25[a]	5% Decalin carboxylic acid-9	47% Pivalic acid 19% C_6, C_7–C_9-acids
Methylcyclopentane	tert. Butanol	15–25	46% 1-Methylcyclopentane-carboxylic acid-1	7% Pivalic acid
Methylcyclohexane	tert. Butanol	15–25	72% 1-Methylcyclohexane-carboxylic acid-1	16% Pivalic acid
Methylcyclohexane	Pentanol-2	15–25	36% 1-Methylcyclohexane-carboxylic acid-1	30% 2,2-Dimethylbutyric acid
Methylcyclohexane	Butadiene-1,3	15–25	15% 1-Methylcyclohexane-carboxylic acid-1	
C_{13}-acid from propene tetramer	tert. Butanol	15–25	6% C_{11}-Dicarboxylic acids	30% Pivalic acid 43% C_{13}-acid
1,4-Dimethylcyclo-hexane carboxylic acid	tert. Butanol	15–25		71% Pivalic acid
2-Methylbutane [641]	tert. Butanol	20	57% 2,2-Dimethylbutyric acid methyl ester 1-methylcyclohexane carboxylic acid (ratio ester to acid 1.4 : 1)	11% Pivalic acid methyl pivalate

[a] Without pressure, formic acid method,
in the experiments with 2-methylbutane and methylcyclohexane methanol was added in the 2nd reaction step

alcohols are converted into the same reaction products as are obtained from the corresponding olefins with the same structure. Differences are only given in the yields obtained.

Generally, higher yields are obtained starting from alcohols, since carbonium ions are formed more slowly and therefore dimerization is suppressed. Alcohols, which are less active, should be converted under pressure. Thus, propanol can be reacted under pressure but no reaction occurs under the conditions of the normal pressure carbonylation with formic acid [591]. Methanol [646, 650] is converted into acetic acid under severe conditions. n-Butanol as well as n-butene react exclusively to give 2-methylbutyric acid [591]. No rearrangement of the n-C_4-carbonium ions to tertiary carbonium ions occurs and therefore no pivalic acid is formed [591]. In the case of pentanols rearrangement is observed to a larger extend [590, 643] (see table 56, page 130).

Tertiary alcohols are converted very easily into carbonium ions. Therefore, they are suitable for carbonylation with formic acid. The reaction of alcohols of the neopentylglycol type is of interest because of their structure which is not comparable to that of olefins. Also these alcohols form no primary carboxylic acids but only tertiary acids by skeletal rearrangement [651].

$$(1)$$

$$(2)$$

$$(3)$$

Ditertiary diols are converted into dicarboxylic acids if the reactive centres are widely separated [1001]. Monocarboxylic acids are obtained exclusively if the branchings are four CH_2-groups distant (see table 61). A distance of 5 to 8 C-atoms favors formation of dicarboxylic acids. An optimum is observed with a distance in the range of 6 or 7 C-atoms [620].

Main products in the carbonylation of primary diols are ω-unsaturated monocarboxylic acids, ω-hydroxy acids and lactones [652, 1001]. However, Schauerte [620] succeeded in carbonylation of decanediol-1,10 to 2,2,7,7-tetramethylsuberic acid together with other products.

Lactones are the main products in the reaction of carbon monoxide and diols with structures that favor lactone formation. Thus, 2,5-dimethyl-hexanediol-2,5 reacts to give the corresponding C_9-lactone [652], whereas e. g. butanediol-1,3 reacts to give tiglic acid in about 10 % yield [653].

Table 61. *Hydrocarboxylation of alcohols and diols via Koch synthesis route*

Starting material	Catalyst	Temp. (°C)	Yield of carboxylic acids (%)	Distribution of reaction products (vol-%)	Ref.
Methanol	BF_3	160–200 (765–1120 atm)	92	Acetic acid	[646, 650]
n-Propanol	H_2SO_4			Isobutyric acid	[591]
Iso-propanol	H_2SO_4			Isobutyric acid	[591]
Butanol-1[a]	H_2SO_4	30	36	85% 2-Methylbutyric acid 15% C_9-acids	[591]
Butanol-2[a]	H_2SO_4	20	43	100% 2-Methylbutyric acid	[591]
tert.-Butanol[a]	H_2SO_4	10–15	78	95% Pivalic acid	[591, 1000]
Pentanol-1[a]	H_2SO_4	10	76	77% 2,2-Dimethylbutyric acid 23% higher acids mainly C_{11}-acids	[591]
Pentanol-2[a]	H_2SO_4	5	81	79% 2,2-Dimethylbutyric acid 21% higher acids mainly C_{11}-acids	[591]
2-Methyl butanol-2[a]	H_2SO_4	5–10	73	10% Pivalic acid 42% 2,2-dimethyl butyric acid 12% C_7-acids 36% higher acids mainly C_{11}-acids	[591]
2,2-Dimethyl-propanol-1[a]	H_2SO_4	22–26	83	100% 2,2-Dimethyl butyric acid	[591]
2,2-Dimethyl-butanol-1[a]	H_2SO_4	10–25	80	63% 2-Methyl-2-ethylbutyric acid 37% 2,2-dimethyl-valeric acid	[591]
2,2-Dimethyl-pentanol[a]	H_2SO_4	28–26	82	65% 2-Methyl-2-ethylvaleric acid 35% 2,2-dimethyl-caproic acid	[591]
1,1-Diethyl-propanol[a]	H_2SO_4	5–10	64	8% C_6–C_7-Acids 45% C_8-Acids 9% C_9–C_{10}-Acids 38% C_{15}-Acids	[591]

Table 61 (continued)

Starting material	Catalyst	Temp. (°C)	Yield of carboxylic acids (%)	Distribution of reactions products (vol- %)	Ref.
2,3,3-Trimethyl-butanol-2[a]	H_2SO_4	10	88	100 % 2,2,3,3-Tetramethyl-butyric acid	[591]
Hexanediol	H_2SO_4			13 % C_8-Dicarboxylic acid 33 % 2,2,4-tri-methylbutyro-lactone	[652]
2,5-Dimethyl-2,6-hexane diol	H_2SO_4	20		44 % C_{10}-Diacid 23.5 C_9-lactone	[652]
1-Hydroxy methyl-1 methylcyclo-pentane[a]	H_2SO_4	19–23	95	1-Methyl cyclo-hexane carboxylic acid-1	[590, 591]
1-Hydroxy methyl-1-methylcyclo-hexane	H_2SO_4	20–25	86	1-Ethyl cyclohexane carboxylic acid	[590, 591]
Hydrindanol-5	H_2SO_4	10–15	56	100 % cis-Hydrin-dane carboxylic acid-8	[643]
β-Decalinol-5	H_2SO_4		95	Decalin-9-carboxylic acid	[590, 654]
cis-8-Hydroxy-methyl hydrindan[a]	H_2SO_4	15–20	68	Decalin carboxylic acid-9	[643]
Decanediol-1,10[a]	H_2SO_4	25	7	2,2,7,7-Tetra-methylsuberic acid	[620]
2,9-Dimethyl-decanediol-2,9[a]	H_2SO_4	0–5	87	63 % 2,2,9,9-Tetramethylsebacic acid	[620]
Pentanol-1	H_2SO_4	20–40 (38–47 atm)	85	58 wt- % 2,2-Dime-thylbutyric acid-methyl ester 27 wt- % 2-ethyl-butyric acid methyl ester	[601, 655]

[a] With formic acid

Koch syntheses are accompanied by isomerization so that in most cases the reaction products are more numerous than the starting material. On the other hand the tendency to undergo isomerization in case of cyclo-aliphatic alcohols favors homogenization of the reaction products. Thus, a mixture of hydrogenated cresols (1,2-, 1,3- and 1,4-cresol) as well as 1-methylcyclohexanol-1 forms only one main product, 1-methyl cyclo-hexane carboxylic acid-1. Similar results are obtained with hydrogenated xylenols.

6.5. Halogenated Compounds

Recently K. E. Möller [600] described the synthesis of halogenated carboxylic acids from halogenated olefins. Thus, methallyl chloride reacted in the presence of BF_3-containing catalysts to give monochloropivalic acid in 70% yield.

$$\underset{\underset{CH_3}{|}}{\overset{\overset{CH_2}{||}}{ClCH_2-C}} \xrightarrow[H^\oplus]{CO/H_2O} \underset{\underset{CH_3}{|}}{\overset{\overset{CH_3}{|}}{ClCH_2-C-COOH}}$$

Other halogenated olefins react analogously.

Table 62 [600]. *Carbonylation of halogenated olefins*

Starting material	Reaction product	Yield (%)
Methallyl chloride	Chloropivalic acid	70
Methallyl bromide	Bromopivalic acid	90
β-Citronellyl chloride	2,2,6-Trimethyl-8-chloro-octanoic acid-1	90
β-Citronellyl bromide	2,2,6-Trimethyl-8-bromo-octanoic acid-1	
8-Bromo octene-1	2,2-Dimethyl-7-bromo-heptanoic acid 2-Methyl-2-ethyl-6-bromo-hexanoic acid	

Carboxylic acids are also formed starting with saturated instead of unsaturated halogenated compounds if severe reaction conditions are applied [591].

Carbonium ions are formed by splitting off hydrogen halide from the allyl halides. Since primary and secondary allyl halogenides are less reactive, only tertiary alkyl halogenides may be converted in high yield. Thus, sec. butyl chloride yields only 30% of 2-methyl butyric acid whereas tert. butyl chloride forms pivalic acid in 76% yield.

6.6. Other Starting Materials

Carboxylic acid esters, N-tert. alkyl acyl amines and saturated aldehydes are also described as starting materials in the Koch carboxylic acid syntheses. Koch and Haaf [591] carbonylated tert. hexylformiate with formic acid and obtained a mixture of carboxylic acids as follows: 15 % C_5- and C_6-acids, 28 % of 2,2-dimethyl valeric acid, 14 % C_8- and C_9-acids and 43 % higher acids, mainly C_{13}-acids. The carbonylation of tert. hexyl acetate yielded 11 % C_5- and C_6-acids, 33 % 2,2-dimethylvaleric acid, 10 % C_8- and C_9-acids and 16 % higher acids, mainly C_{13}-acids.

7. Industrial Application and Economic Aspects

After a license had been granted by Koch to the Royal Dutch Shell Laboratory at Amsterdam, the synthesis was developed by Shell for industrial application. The first unit for continous production was built by Shell at Pernis near Amsterdam. The so-called Versatic acids have been

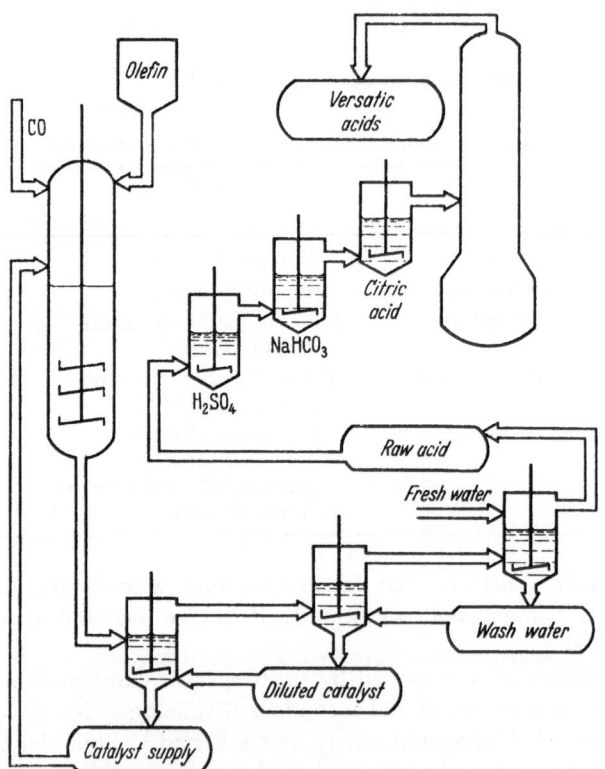

Fig. 19. Flow sheet of Versatic acid plant

manufactured for some years in continuous operation. Production was started with a 5000-ton annual capacity. Fig. 19 gives a simplified flow sheet of the Shell Versatic acid unit.

Especially isobutylene and diisobutylene are converted into pivalic acid and olefin cuts in the range of C_6 to C_{10} into C_7 to C_{11}-carboxylic acid mixtures. This process employs CO pressures of about 70 atm and temperatures of 70 °C in the presence of H_3PO_4/BF_3 as catalyst. Over 90 % tertiary acids are obtained.

Another plant licensed by H. Koch to Enjay Chemical Co. in Baton Rouge, USA, came onstream in 1965. This plant is designed to make the so-called Neo-acids (branched C_5-, C_7- and C_{10}-acids) with a capacity of 4500 tons per year. As catalyst $BF_3/2H_2O$ is used [796, 1002].

An elegant technique carried out on a pilot plant scale was recently reported by Pawlenko of Schering AG. [999], see fig. 20.

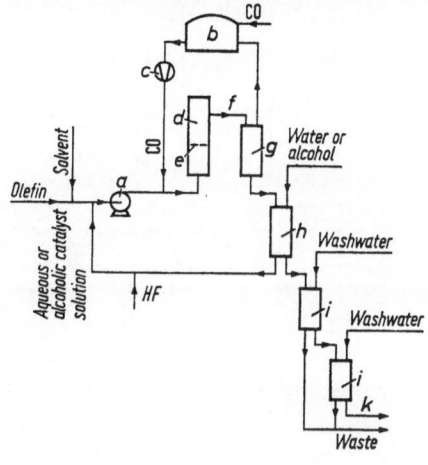

a	Liquid pump	*f*	Liquid overflow
b	CO-gasmeter	*g*	Pressure separator
c	Gas compressor	*h*	Pre washer
d	Pressure reactor	*i*	Neutral washer
e	Baffle	*k*	Product (acid or ester)

Fig. 20

In this method the heavy catalyst phase remains in the reactor. Olefin plus solvent are fed to the bottom and the product formed, which is insoluble in the catalyst phase, is taken overhead together with the solvent.

The acids available by Koch syntheses are of industrial interest [1002] because of their chemical configuration which is characterized by excellent hydrolytic and thermal stability of their esters [664–667]. These esters are used to make resins and varnishes.

Finally chloropivalic acid obtained by K. E. Möller from methallyl chloride is of interest as intermediate in the α,α-dimethyl-β-propiolactone synthesis. This lactone shows promise as a monomer for polyester fibers. Studies for polymerization of α,α-dimethyl-β-propiolactone are under way at several companies (Shell, Eastman, Kodak, ICI, Montecatini [424, 432, 510, 794, 795]).

In 1965/66 the capacity of DuPont glycolic acid production was estimated at 60,000 tons per year.

IV

Ring Closures with Carbon Monoxide

1. General Considerations

Unsaturated compounds containing a nucleophilic group and a reactive hydrogen atom in a position which favors ring closure may react with carbon monoxide to give five- or six-membered rings [668] (1) (2).

(1) 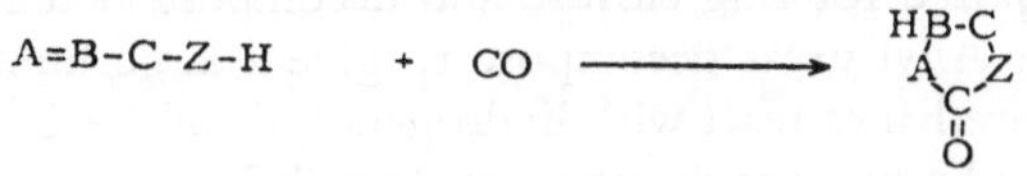   A=B–C–Z–H   +   CO $\longrightarrow$
$$\begin{array}{c} \text{HB--C} \\ \text{A}\diagdown\quad\diagup\text{Z} \\ \text{C} \\ \| \\ \text{O} \end{array}$$

(2) 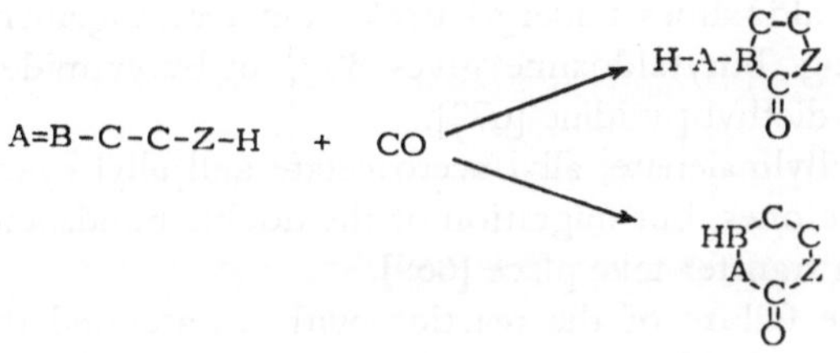   A=B–C–C–Z–H   +   CO
$$\begin{array}{c} \text{C--C} \\ \text{H--A--B}\diagdown\quad\diagup\text{Z} \\ \text{C} \\ \| \\ \text{O} \end{array}$$
$$\begin{array}{c} \text{HB}\diagdown\text{C--C} \\ \text{A}\diagup\quad\diagdown\text{Z} \\ \text{C} \\ \| \\ \text{O} \end{array}$$

Thus, carbon monoxide reacts with unsaturated amides to give imides, with unsaturated amines to give lactams, with unsaturated alcohols to give lactones, with Schiff bases, aromatic ketoximes, phenylhydrazones, semicarbazones, azines and nitriles to give phthalimidines, with azobenzenes to give indazolones or 2,4-dioxo-1,2,3,4-tetrahydroquinazolines, with cumulenes to give indones and with dienes to give unsaturated ketones.

The formation of phenols, which is a side reaction in the formation of doubly unsaturated acids from allyl halides, acetylene and carbon monoxide, as described by Chiusoli, may also be explained by the above scheme.

A number of starting materials having structures like those given in equations (1) and (2) do not undergo ring closure. Thiols, well known for their thermal instability, mainly polymerize and give only traces of thiolactones [669] besides a number of other compounds [670]. Like other sulfur compounds they also form complexes with the carbonyl catalyst [135–144]. Acrylic and methacrylic acid polymerize under the reaction conditions generally applied in the ring closure reaction. Long-chain α,β unsaturated acids yield lactones through migration of the double bond and intramolecular addition of the carboxyl group to the double bond in its new position.

$H_3C-CH_2-CH_2-CH=CH-COOH$

$\updownarrow$

$H_3C-CH_2-CH=CH-CH_2-COOH \rightleftharpoons H_3C-CH=CH-CH_2-CH_2-COOH$

Unsaturated sulfonamides are recovered unchanged at temperatures normally required for ring closure and decompose at temperatures above 300 °C [669]. Allyl ureas decompose to give a large number of products [671]. Allyl urethanes react with hydrogenolysis of the C−N bond to form carbamates and a number of other products [671].

Aliphatic aldoximes undergo Beckmann rearrangement to form carboxamides, e. g. butyraldoxime gives 38% of butyramide besides 23% of 2-propyl-3,5-diethyl-pyridine [672].

Diethyl allylmalenate, allyl acetoacetate and allyl cyanoacetate do not give cyclic ketones, but migration of the double bonds and hydrogenation by hydrogen transfer take place [669].

While the failure of the reaction with unsaturated thiols, acids, urethanes, ureas and aldoximes must be attributed to side reactions, with the three allyl compounds mentioned above it is rather due to the decreased electron density at the atom Z which is too low for a nucleophilic attack on the acylium cation (see next chapter on the mechanism of the ring closure reaction).

As can be seen from the examples given on the following pages, ring closure is catalyzed by metal carbonyls. The reaction is not limited to C=C double bonds but can also be achieved with C=N and N=N double bonds.

2. Reaction Mechanism

The mechanism of ring closure is closely related to the mechanism of the hydroformylation reaction. Despite the absence of sufficient experimental evidence to support a rigorous mechansism, it may be assumed that the reaction will proceed through steps (1)–(3) [668].

It is assumed that in the first step the metal hydrocarbonyl is added across the double bond of the starting material, analogous to the hydroformylation reaction (1).

$$A=B-\overset{\underset{|}{R}}{C}-\underline{Z}-H \xrightarrow{+\ HCo(CO)_4} A-B-\overset{\underset{|}{H}}{C}-\overset{\underset{|}{R}}{\underline{Z}}-H \qquad (1)$$
$$\underset{Co(CO)_4}{}$$

$$A-B-\overset{\underset{|}{H}}{\underset{Co(CO)_4}{C}}-\underline{Z}-H \xrightarrow{+CO} A-B-\overset{\underset{|}{H}}{C}-\overset{\underset{|}{R}}{\underline{Z}}-H \qquad (2)$$
$$\overset{\oplus}{C}=O$$
$$\overset{\ominus}{Co(CO)_4}$$

$$A-B-\overset{\underset{|}{H}}{C}-\underline{Z}-H \longrightarrow \left[\begin{array}{c} \overset{H}{\underset{A}{B}}-C-\\ \overset{\oplus}{C}\overset{Z-R}{\underset{H}{}}\\ O\\ Co(CO)_4^{\ominus} \end{array}\right] \xrightarrow{-HCo(CO)_4} \begin{array}{c} B-C \\ A \quad Z' \\ C \quad R \\ O \end{array} \qquad (3)$$
$$\overset{\oplus}{C}=O$$
$$\overset{\ominus}{Co(CO)_4}$$

The following facts support the assumption that the true catalysts are the transition metal hydrocarbonyls rather than the corresponding metal carbonyls. Only metals which can form hydrocarbonyls, like cobalt, rhodium and iron, can act as catalysts [123, 280, 673, 674], whereas nickel, e.g., is inactive [123, 673] in most cases. Furthermore it is known that cobalt, rhodium and iron carbonyls, under the reaction conditions applied, are able to abstract hydrogen from alcohols, amines and even from the unreactive cycloparaffins to form metal hydrocarbonyls [121–124].

Moreover it was shown that unsaturated amides react already at −30 to −20 °C with stoichiometric amounts of cobalt hydrocarbonyl with ring closure to give imides, whereas no reaction occurs under these conditions with $Co_2(CO)_8$ [668].

The reaction illustrated in (1) is probably a sequence of three steps (1a)–(1c).

$$HCo(CO)_4 \rightleftharpoons HCo(CO)_3 + CO \qquad (1a)$$

$$A=B-\overset{\underset{|}{R}}{C}-\underline{Z}-H + HCo(CO)_3 \longrightarrow A=B-\overset{\underset{|}{R}}{C}-\underline{Z}-H \qquad (1b)$$
$$\underset{(CO)_3CoH}{}$$

$$\underset{(CO)_3CoH}{A=B-\overset{\underset{|}{R}}{C}-\underline{Z}-H} \longrightarrow A-BH-\overset{\underset{|}{R}}{\underset{Co(CO)_3}{C}}-\underline{Z}-H \xrightarrow{CO} A-BH-\overset{\underset{|}{R}}{\underset{Co(CO)_4}{C}}-\underline{Z}-H \qquad (1c)$$

The coordinatively unsaturated cobalt hydrotricarbonyl which is formed via (1a) reacts with the olefinic compound with formation of a π-complex (1b), the π-complex is rearranged with formation of a cobalt-carbon

σ-bond to yield an alkyl cobalt tricarbonyl which is coordinatively saturated by reaction with one more molecule of carbon monoxide to give an
alkylcobalttetracarbonyl (1 c).

Cobalt hydrocarbonyl reacts like a hydride when adding across the
C=C double bond in (1c). This explains why its reaction with unsaturated
amines or alcohols in aromatic solvents leads preferentially to compounds
in which the metal is bound in the terminal position.

The addition of metal hydrocarbonyls to Schiff bases, oximes, azo compounds, phenylhydrazones, semicarbazones, and azines can easily be explained on the basis of the polar structures. If the electron density on the C-
atom of the C=N double bond in Schiff bases is increased by a methyl
substituent, the yield decreases. If the electron density is lowered by a
phenyl substituent, yields are remarkably increased.

The electron-attracting carbamoyl group in unsaturated carboxamides
in analogy to the carboxyl group of unsaturated esters should direct the
hydrocarbonyl addition such that an α-metal substituted compound is
formed (4). It is known that unsaturated esters also react with hydrocarbonyl at low temperatures (50 to 70 °C) to give products with Co attached

$$
\underset{\substack{|\\ }}{\overset{\substack{|\quad|\quad O\\ |\quad|\quad\|}}{C=C-C}}-NHR \xrightarrow{+\,HCo(CO)_4} \underset{\substack{|\quad|\\ \quad Co(CO)_4}}{\overset{\substack{|\quad|\quad O\\ |\quad|\quad\|}}{HC-C-C}}-NHR \qquad (4)
$$

$$
\xrightarrow{\;\Delta\;} \underset{\substack{|\quad|\\ Co(CO)_4}}{\overset{\substack{|\quad|\quad O\\ |\quad|\quad\|}}{HC-C-C}}-NHR \xrightarrow{+\,CO} \underset{\substack{|\quad|\\ C=O\\ |\\ Co(CO)_4}}{\overset{\substack{|\quad|\quad O\\ |\quad|\quad\|}}{HC-C-C}}-NHR \qquad (5)
$$

$$
\longrightarrow \quad \underset{[Co(CO)_4]^{\ominus}\;\overset{|}{\underset{H}{N}}-R}{O=C^{\oplus}\diagdown C=O} \xrightarrow{-HCo(CO)_4} \underset{R}{O\diagdown N\diagup O}
$$

to the α-carbon-atom [35, 60, 61]; these products react with CO and hydrogen or alcohols to give the corresponding α-formyl carboxylic acid esters
or the malonic acid diesters, respectively. However, at higher temperatures (110 to 150 °C) the same starting materials lead to β-substituted carboxylic acid esters [60, 675]. Compounds with longer alkyl chains give
products containing even ω-formyl-substituted compounds, formed through
isomerization of the intermediate cobalt complexes. The metal substituent
migrates towards the end of the chain in these isomerizations. Obviously
unsaturated amides behave analogously (4) (5). These isomerizations are

closely related to those observed in hydroformylation (see page 5) and with alkylboron [71, 76], alkylsilicon [77, 79], and alkylaluminium compounds [80] (for theoretical explanations see page 6).

However, it should be noted at this stage of the discussion that the metal-alkyl compounds resulting from the addition of the hydrocarbonyls to the unsaturated compounds must remain hypothetical. No such alkyl-cobalt compounds have been isolated or definitely proved so far [35, 37].

In papers on the mechanism of hydroformylation, it has been suggested that alkyltetracarbonylcobalt compounds may rearrange to acyltricarbonyl-cobalt compounds followed by formation of $HCo(CO)_3$ by reaction with hydrogen or alcohols without CO being taken up from the gas phase [35] (see Eq. 6).

$HCo(CO)_3$ could then either react again with an olefinic compound, or could take up carbon monoxide from the gas phase to form cobalt hydro-tetracarbonyl [35, 37].

$$
\begin{array}{ccc}
\underset{\underset{Co(CO)_4}{|}}{C-C-C} & \longrightarrow & \underset{\underset{\underset{Co(CO)_3}{|}}{C=O}}{C-C-C} \xrightarrow[-HCo(CO)_3]{+H_2} \underset{\underset{CHO}{|}}{C-C-C} \qquad (6)
\end{array}
$$

Although a similar mechanism in the ring closure reaction cannot be excluded, it is more likely, in view of the high CO pressure applied, that the reaction proceeds via (5).

The last step in the reaction is the ring closure with simultaneous regeneration of cobalt hydrotetracarbonyl. The ring is closed by a nucleophilic attack of atom Z on the acyl cation with simultaneous hydrogen transfer from Z to the $(Co(CO)_4)^-$, resp. $Co(CO)_3^-$, anion.

It is easy to see from the experimental results how the ring closure is influenced by the electron densitiy on the atom Z. Thus, for example, better yields are obtained with methylallylamine or N-methylacrylamide than with their phenyl homologues [673, 674].

Similarly, higher yields are obtained from Schiff bases derived from p- or m-dimethylaminobenzaldehyde and from p- or m-methoxybenzaldehyde, or the analogous azo compounds [123], than from the unsubstituted compounds; this is certainly due to inductive effects of the substitutents resulting in increased electron density in the aromatic ring.

Only one result appears not to fit in the proposed mechanism: the reaction of the Schiff base of 2-naphthaldehyde [123]. In this case the ring closure takes place, not on C-1, as expected, but on C-3; however, in this example special steric effects are involved which will be discussed later (see page 160).

A mechanism proposed by S. Horiie and S. Murahashi [676] for the carbonylation of Schiff bases and azo compounds, and by Rosenthal [122] for the carbonylation of phenylhydrazones, has been shown to be incorrect. $Co_2(CO)_8$ was thought to react with the lone electron pair on the nitrogen atom to form a π-compound; insertion of carbon monoxide should then lead to an unsaturated Co-acyl compound containing 4-coordinate nitrogen, which subsequently should undergo ring closure in a multi-center reaction, the aromatic hydrogen in the o-position being simultaneously transferred to the C=N or C=C double bond. However, Rosenthal and Gervay [677] later showed that in the reaction of nitriles with CO, H_2 and D_2, cyclization is not accompanied by simultaneous transfer of the o-hydrogen to the double bond.

In a recent publication [797] Rosenthal and Wender also favor the mechanism described earlier in this chapter, previously proposed by J. Falbe [668].

3. Catalysts, Reaction Conditions and Solvents

Suitable catalysts for ring closure reactions are cobalt carbonyls [123, 280, 673, 674], rhodium carbonyls [280, 678], iron carbonyl [123] and certain palladium compounds [679]. Nickel carbonyls, the active catalysts in the Reppe syntheses, are inactive in most cases [123, 673]. A few examples in which nickel is active are the formation of phenols from allyl halides, acetylene and carbon monoxide, which is only a side reaction, and the mechanistically unclear formation of lactones from allyl carbinol and butyne-1-ol-4 [438].

Ring closure reactions proceed under pressure (100 to 300 atm) and at temperatures of 120 to 300 °C, preferably between 170 to 280 °C.

Suitable solvents are aromatic or aliphatic hydrocarbons. Cyclic ethers such as tetrahydrofuran or dioxane may be used where the starting materials or the reaction products are highly polar.

4. Special Ring Closures with Carbon Monoxide

4.1. Imides from Unsaturated Amides

Unsaturated amides react with carbon monoxide in the presence of cobalt catalysts to give imides [673, 681–683]. Acrylamide reacts with carbon monoxide to give succinimide in 82% yield. N-mono-substituted acrylamides are converted into the corresponding N-substituted succinimides (see table 63).

$$H_2C=CH-\underset{\underset{O}{\|}}{C}-NHR + CO \xrightarrow{Co_2(CO)_8}$$

$$R = H \qquad\qquad\qquad R = H$$
$$R \neq H \qquad\qquad\qquad R \neq H$$

When the starting materials contain halogenated aromatic groups, no dehalogenation takes place. Similarly, methacrylamide and N-substituted methacrylamides give α-methylsuccinimide (see table 63).

β-Arylated acrylamides also react to give α-substituted succinimides. Thus α-phenylsuccinimide is obtained from cinnamamide.

$$C_6H_5-HC=CH-\underset{\underset{O}{\|}}{C}-NH_2 + CO \xrightarrow{Co_2(CO)_8}$$

The reaction of β-alkylacrylamides leads not only to α-alkylsuccinimides, but also to glutarimides. For example, crotonamide reacts to form a mix-

$$H_3C-\underset{\underset{R}{|}}{C}=CH-\underset{\underset{O}{\|}}{C}-NH_2 + CO \xrightarrow{Co_2(CO)_8}$$

$$R = H \qquad\qquad 68\% \qquad\qquad 19\%$$
$$R = CH_3 \qquad\qquad\qquad\qquad 67\%$$

ture of α-methylsuccinimide and glutarimide and β,β-dimethylacrylamide gives only β-methylglutarimide.

The six-membered rings are formed by isomerization of intermediate complexes formed from the olefinic starting material and the catalyst. Similar isomerizations were observed in other reactions with carbon monoxide in the presence of transition metal carbonyls (see the chapter on hydroformylation and carbonylation reactions).

The exclusive formation of the six-membered cyclic imide from β,β-dimethylacrylamide has a parallel in the oxo synthesis, where no quaternary

Table 63. *Syntheses of succinimides from acrylamides*

Starting material	Reaction product	Yield (%)	Ref.
Acrylamide	Succinimide	82	[673]
N-Methylacrylamide	N-Methylsuccinimide	94	[671, 673]
N-Butylacrylamide	N-Butylsuccinimide	72	[673]
N-Isobutylacrylamide	N-Isobutylsuccinimide	80	[673]
N-Hexylacrylamide	N-Hexylsuccinimide	77	[673]
N-Dodecylacrylamide	N-Dodecylsuccinimide	85	[671, 673, 681]
N-Acrylo-ethylglycinate	N-Succinyl-ethylglycinate	70	[673]
N-Phenylacrylamide	N-Phenylsuccinimide	64	[673]
N-(p-Chlorophenyl)acrylamide	N-(p-Chlorophenyl)-succinimide	65	[673]
N-(2,6-Dichlorophenyl) acrylamide	N-(2,6-Dichlorophenyl)-succinimide	44	[673]
N-Benzylacrylamide	N-Benzylsuccinimide	92	[673]
N-Allylacrylamide	N-Allylsuccinimide	55	[671, 674]
Methacrylamide	α-Methylsuccinimide	68	[673]
N-Methylmethacrylamide	N-Methyl-α-methyl-succinimide	70	[673]
N-n-Butylmethacrylamide	N-n-Butyl-α-methyl-succinimide	74	[673]
N-Benzylmethacrylamide	N-Benzyl-α-methylsuccinimide	76	[673]
Cinnamamide	α-Phenylsuccinimide	32	[673]

carbon atoms are formed in the presence of cobalt catalysts [25]. Thus the hydroformylation of isobutene gives almost exclusively isopentanol with only small quantities of neopentanol [230] and 2,6-dimethyl-5,6-dihydro-4 H-pyran gives only products with the formyl group in the 3-position, whereas 5,6-dihydro-4 H-pyran gives mainly products with the formyl group attached to the 2-position [283].

β,γ-Unsaturated amides also react in accordance with the above scheme and give glutarimides. Unsaturated alicyclic amides yield five-membered glutarimides.

$$R-CH=CH-\underset{\underset{CH_3O}{|}}{\overset{\overset{CH_3}{|}}{C}}-\underset{\overset{\|}{O}}{C}-NH_2 + CO \xrightarrow{Co_2(CO)_8}$$

R = H   58%
R = CH$_3$ 64%

Unsaturated alicyclic amides yield five-membered or six-membered bicyclic imides, depending on the position of the double bond.

Unsaturated N,N-dialkylamides and aromatic amides, such as benzamide, do not react to give imides.

4.2. Lactams from Unsaturated Amines

Unsaturated amines react with carbon monoxide in the presence of cobalt catalysts to give five- or six-membered lactams.

$$H_2C=CH-CH_2-NH_2 + CO \longrightarrow$$

The N-alkylallylamines which are readily accessible from allyl chloride and primary amines or from allylamine and alkyl chlorides [684] react to form N-alkyl-2-pyrrolidones.

$$H_2C=CH-CH_2-NHR + CO \longrightarrow$$

Table 64. *Reaction of N-substituted allylamines with carbon monoxide in the presence of* $Co_2(CO)_8$

Starting material	Reaction product	Yield (%)	Ref.
N-Methylallylamine	N-Methylpyrrolidone	78	[674]
N-Ethylallylamine	N-Ethylpyrrolidone	61	[674]
N-Isobutylallylamine	N-Isobutylpyrrolidone	61	[674]
N-Octylallylamine	N-Octylpyrrolidone	58	[674]
N-Dodecylallylamine	N-Dodecylpyrrolidone	47	[674]
N-Phenylallylamine	N-Phenylpyrrolidone	26	[681]

The carbonylation of diallylamine leads to a mixture of N-allyl-2-pyrrolidone, N-propenyl-2-pyrrolidone, and 2-pyrrolidone.

The principal product may be any one of the three lactams, depending on the reaction conditions. The migration of the double bond again is probably due to isomerization promoted by the metal carbonyl catalyst. The formation of 2-pyrrolidone, which becomes appreciable only at high temperature, is accompanied by formation of propylene. Under these conditions, the C–N-bond evidently undergoes hydrogenolytic cleavage with hydrogen transfer.

Carbonylation of N-alkylated β-methylallylamines yields 1-alkyl-4-methyl-2-pyrrolidones.

γ-Alkylallylamines give mixtures of 3-alkyl-2-pyrrolidones and 2-piperidones.

As in the synthesis of imides, the formation of six-membered lactam rings is attributed to the isomerization of intermediate complexes of the catalyst and the starting material.

Isomerizations also occur during the carbonylation of unsaturated alicyclic amines; for example, the carbonylation of (3-cyclohexenyl-methyl) amine leads, besides the expected bicyclic six-membered lactam, also to a five-membered lactam.

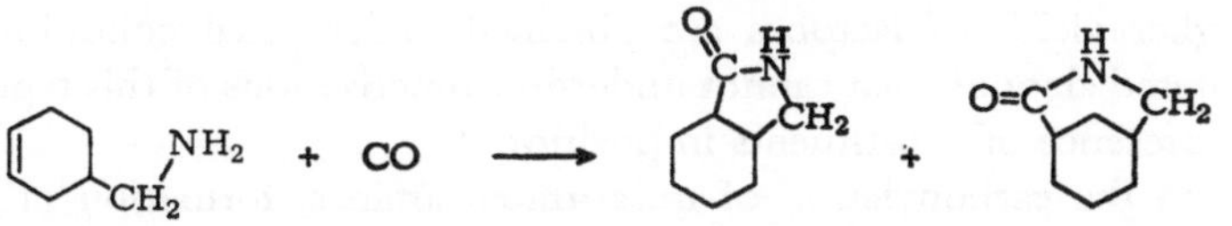

N-Alkenylformamides are formed only in small amounts in all the above reactions.

4.3. Lactones from Unsaturated Alcohols

Unsaturated primary alcohols and carbon monoxide react with ring closure to give lactones [280].

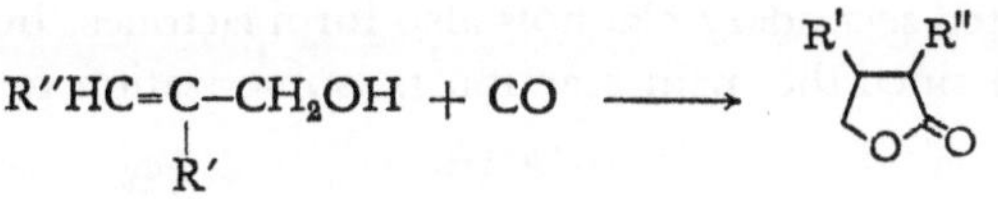

$$R''HC=C-CH_2OH + CO \longrightarrow$$

$$\underset{R'}{|}$$

$$R' = R'' = H$$
$$R' = CH_3 \quad R'' = H$$
$$R' = H \quad R'' = CH_3$$

However, this is often accompanied by an undesirable side-reaction, namely the carbonyl-catalyzed isomerization of the unsaturated alcohols to saturated aldehydes.

$$H_2C=CH-CH_2OH \longrightarrow [CH_3-CH=CHOH] \longrightarrow H_3C-CH_2-CHO$$

Thus allyl alcohol, methallyl alcohol and crotyl alcohol give only about 2% of the γ-lactones under the reaction conditions usually applied in the ring closure reaction. However, the selectivity towards lactones is increased with the use of acetonitrile or its derivatives as solvents and with addition of small amounts of organic bases such as pyridines in combination with acetonitrile. A 60% yield of γ-butyrolactone may be obtained under these conditions [835], while the rest of the starting materials isomerize to the corresponding aldehydes. These undergo further reactions such as the aldol condensation, or polymerize at higher temperatures.

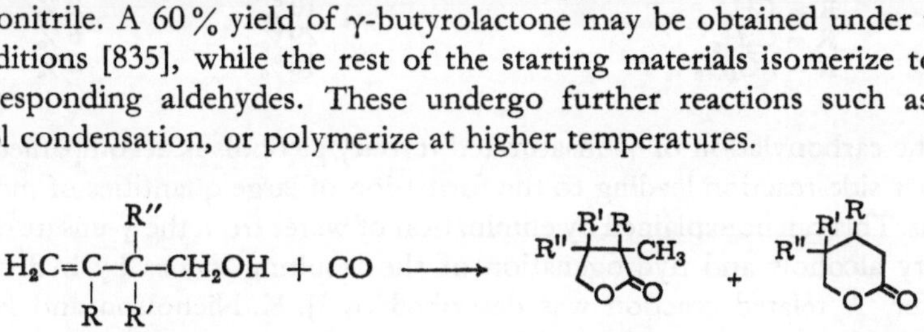

$$\underset{R \ R'}{\overset{R''}{H_2C=C-C-CH_2OH}} + CO \longrightarrow$$

R = H R' = R'' = CH₃	51%	14%
R = R' = R'' = CH₃	3%	25%
R = H R' = CH₃ R'' = C₂H₅	40%	13%

Higher yields of lactones are obtained under usual conditions from unsaturated alcohols that cannot undergo isomerizations of this type owing to the presence of substituents in position 2.

As in the carbonylation of unsaturated amines, formation of the six-membered ring is favored by an alkyl substituent in position 3.

Unsaturated alicyclic alcohols also give lactones. The reaction of (3-cyclohexenyl)methanol leads not to the expected bicyclic δ-lactone but to the isomeric bicyclic γ-lactone, probably again by isomerization of an intermediate cobalt complex.

γ-Unsaturated secondary alcohols also form lactones, but only in small yields, ($\sim 2\%$) since the main reaction is isomerization to the ketones.

$$H_2C=CH-CH_2-\underset{\underset{OH}{|}}{CH}-R + CO \longrightarrow$$

$$R = C_2H_5$$
$$R = C_3H_7$$

$$H_3C-CH_2-CH_2-\underset{\underset{O}{\|}}{C}-R$$

Since γ-unsaturated tertiary alcohols cannot isomerize to carbonyl compounds, they generally give better yields of lactones.

$$H_2C=CH-CH_2-\underset{\underset{CH_3}{|}}{\overset{\overset{R}{|}}{C}}-OH + CO \longrightarrow$$

R		
R = CH$_3$	10%	2%
R = C$_2$H$_5$	29%	6%
R = i-C$_4$H$_9$	10%	2%

The carbonylation of γ-unsaturated tertiary alcohols is accompanied by another side-reaction leading to the formation of large quantities of mono-olefins. This can be explained by elimination of water from the γ-unsaturated tertiary alcohols and hydrogenation of the resulting dienes by hydrogen transfer. A related reaction was described by J. K. Nicholson and B. L. Shaw [685], who obtained propylene and acrolein when an aqueous solution of allyl alcohol was heated in the presence of catalytic amounts of RuCl$_3$.

$$\text{H}_2\text{C=CH–CH}_2\text{–}\overset{\displaystyle \text{CH}_3}{\underset{\displaystyle \text{CH}_3}{\text{C}}}\text{–OH}$$

$$\overset{\displaystyle \text{CH}_3}{\underset{\displaystyle \text{CH}_3}{\text{H}_3\text{C–CH=CH–CH}}} \qquad \overset{\displaystyle \text{CH}_3}{\underset{\displaystyle \text{CH}_2}{\text{H}_3\text{C–CH}_2\text{–CH}_2\text{–C}}} \qquad \overset{\displaystyle \text{CH}_3}{\underset{\displaystyle \text{CH}_3}{\text{H}_3\text{C–CH}_2\text{–CH=C}}}$$

4.4. Phthalimidines from Schiff Bases or Aromatic Nitriles

The ring-closure reaction is not confined to compounds containing a
C=C double bond, and the atom represented by Z in equation (1) page 147
needs not be a heteroatom. For example, Schiff bases, in which the unsaturated bond is a C=N double bond and Z is a carbon atom, react with carbon
monoxide to give high yields of phthalimidines [123, 676, 686].

Table 65. *Reaction of Schiff bases with CO in the presence of $Co_2(CO)_8$
to yield phthalimidines*

R'	R"	R	Phthalimidine	Yield (%)	Ref.
H	C_6H_5	H	2-C_6H_5	84	[686]
H	p-$CH_3OC_6H_4$	H	2-p-$CH_3OC_6H_4$	86	[123]
H	p-$OH-C_6H_4$	H	2-p-HOC_6H_4	65	[123]
H	p-ClC_6H_4	H	2-p-ClC_6H_4	75	[123]
H	p-$(C_2H_5)_2NC_6H_4$	H	2-p-$(C_2H_5)_2NC_6H_4$	—	[686]
H	1-naphthyl	H	2-(1-naphthyl)	55	[686]
p-$(CH_3)_2N$	C_6H_5	H	2-C_6H_5, 6-$(CH_3)_2N$	82	[123]
p-HO	C_6H_5	H	2-C_6H_5, 6-HO	77	[123]
p-CH_3O	C_6H_5	H	2-C_6H_5, 6-CH_3O	86	[686]
p-Cl	C_6H_5	H	2-C_6H_5, 6-Cl	45	[686]
m-$(CH_3)_2N$	C_6H_5	H	2-C_6H_5, 7-$(CH_3)_2N$	87	[687]
H	C_6H_5	CH_3	2-C_6H_5, 3-CH_3	61	[123,686]
H	C_6H_5	C_6H_5	2-C_6H_5, 3-C_6H_5	97	[123,686]
o-CH_3O	C_6H_5	H	2-C_6H_5, 4-CH_3O	18	[123]
m-CH_3O	C_6H_5	H	2-C_6H_5, 5-CH_3O	5	[123]
m-HO	C_6H_5	H	2-C_6H_5, 7-HO	77	[687]

N-Benzylidenaniline gives 2-phenylphthalimidine in 84% yield. N-Benzylidenanilines with o,p-directing substituents also undergo this reaction.

If, on the other hand, the starting compound contains a nitro group, the reaction fails. It appears that the nitro group is reduced, and the product then undergoes further reactions. Lower yields are obtained if the aromatic ring of the benzylidene group in N-benzylidenaniline carries an o,p-directing substituent, such as a hydroxy or methoxy group, in the o-position (table 65).

Schiff bases of aliphatic amines also give phthalimidines.

$$R = CH_3 \qquad\qquad\qquad 79\%$$
$$R = CH_2\text{-}C_6H_5 \qquad\qquad 82\%$$

Compounds containing two phthalimidine groups can also be prepared in this way.

49%

Schiff bases from naphthaldehydes react in a similar manner. The reaction of 1-naphthalidenaniline leads to N-phenylbenzo(e)phthalimidine. Substitution in position 8 was not observed.

2-Naphthalidenaniline gives, not N-phenylbenzo(g)phthalimidine, but only N-phenylbenzo(f)phthalimidine. On the basis of the charge distribution, cyclization would be expected in position 1 as discussed in the chapter on reaction mechanism.

However, Stuart-Briegleb molecular models show that ring closure in this position is subject to strong steric hindrance, whereas ring closure in position 3 is unhindered.

The synthesis of N-benzylphthalimidine by reaction of benzonitrile [688] with carbon monoxide or synthesis gas in the presence of cobalt carbonyls may be regarded as a special case of the reaction with Schiff bases. It is known that nitriles can give Schiff bases as intermediates on hydrogenation [689]. From the products obtained in the carbonylation of nitriles, and from the fact that higher yields are obtained with synthesis gas than with pure carbon monoxide [677], as well as from the observation that the best yields are obtained when 1 mole of benzylamine is added per mole of benzonitrile [677], it is very probable that the formation of the phthalimidine in this case again proceeds via the Schiff base.

22%

4.5. Phthalimidines from Aromatic Ketoximes, Phenylhydrazones, Semicarbazones and Azines

Aromatic ketoximes react with a CO/H_2 mixture (98.5 : 1.5) to give phthalimidines [690]. The initial products are probably N-hydroxyphthalimidines which are then hydrogenated to phthalimidines.

R = phenyl, R′ = H   R = phenyl R′ = H 80%
R = phenyl, R′ = COOH   R = phenyl R′ = H 86%
                               (R′ = H after decarboxylation)
R = CH₃   R′ = H   R = CH₃   R′ = H
R = benzyl   R′ = H   R = benzyl R′ = H

Reactions with alkyl aryl ketoximes are less selective than reactions with diaryl ketoximes.

Naphthketoximes also react. Thus methyl 2-naphthylketoxime gives a phthalimidine together with other products.

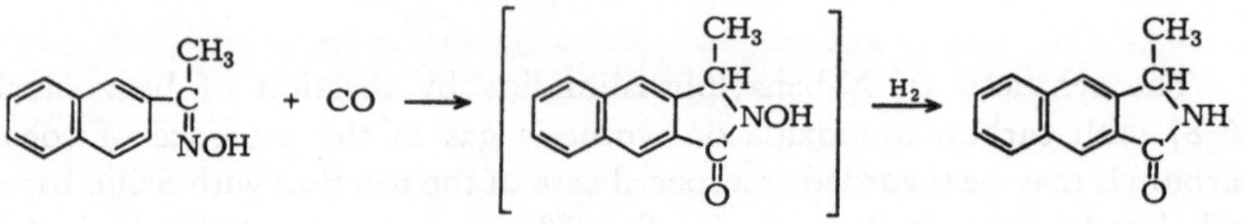

10%

As in the case of 2-naphthalidenaniline, ring closure again takes place in position 3 because of steric hindrance. Attempts to prepare the N-hydroxy- or N-methoxy-methylphthalimidines were unsuccessful [690]; the products in every case were simply phthalimidines. The N−O bond appears to be very susceptible to hydrogenolysis by the hydrogen present.

Phenylhydrazones of diaryl- or aryl-alkyl-ketones also undergo ring closure with carbon monoxide [122, 691, 692], however, the expected N-phenyl aminophthalimidines cannot be isolated. As in the experiments with aromatic oximes, the N–N bond suffers hydrogenolysis at 190 to 200 °C. At higher temperatures, the hydrogenolysis gives way to the insertion of carbon monoxide into the N–N bond, leading to formation of N-(N'-phenylcarbamoyl)-phthalimidines.

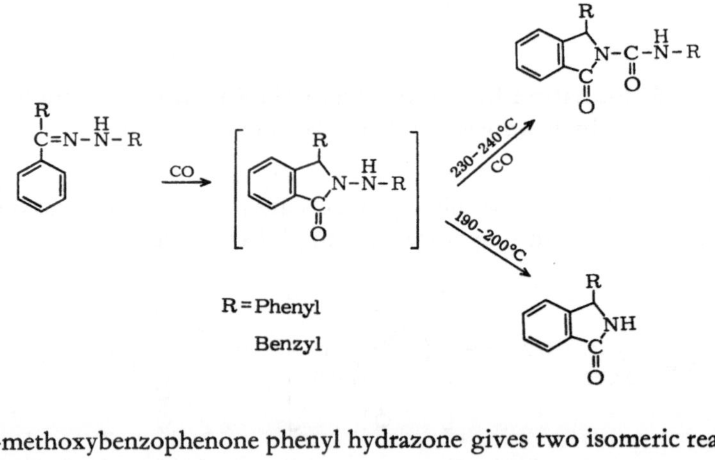

R = Phenyl

Benzyl

4-methoxybenzophenone phenyl hydrazone gives two isomeric reaction products.

Thus ring closure does not take place exclusively on the substituted ring, as in the case with Schiff bases and with the azo compounds described below, but proceeds to an equal extent on the unsubstituted ring.

Table 66. *Carbonylation of phenylhydrazones of diaryl- and aryl-alkyl-ketones in the presence of $Co_2(CO)_8$*

Starting material	Reaction product	Temp. (°C)	Yield (%)	Ref.
Benzophenone phenyl-hydrazone	3-Phenylphthal-imidine	190–200	25	[691]
Benzophenone phenyl-hydrazone	3-Phenylphthal-imidine-N-carbox-anilide	210–220	12	[691]
	3-phenylphthal-imidine		50	
Benzophenone phenyl-hydrazone	3-Phenylphthal-imidine-N-carbox-anilide	230–240	70	[691]
4-Methylbenzophenone phenylhydrazone	3-(p-Tolyl)-phthal-imidine-N-carbox-anilide	230	20	[691]
	3-phenyl-6-methyl-phthalimidine-N-carboxanilide		20	
Acetophenone phenyl-hydrazone	3-Methyl-phthalimidine-N-carboxanilide	230		
	3-methyl-N-phenyl phthalimidine			[691]
Desoxybenzoin phenyl-hydrazone	3-Benzylphthalimidine n-carboxanilide	235	22	[692]
	3-benzyl-2-phenyl-phthalimidine		11	

Phenylhydrazones of aromatic aldehydes also give phthalimidines [122, 691]. With the use of compounds labelled with ^{15}N it was shown that the nitrogen originally attached directly to the phenyl group of the phenyl-hydrazone is retained in the indazolone.

Whereas the C=N double bond in the phenylhydrazones of aromatic ketones reacts with its original position, owing to conjugation with the two aromatic rings, the phenylhydrazones of aromatic aldehydes, like those

of aliphatic aldehydes and ketones [694], exist partly in the tautomeric azo form. The phenylhydrazones of aromatic aldehydes evidently react in the azo form with carbon monoxide. However, the expected six-membered heterocycles are not obtained; one nitrogen atom is eliminated as ammonia.

Table 67. *Carbonylation of phenylhydrazones of aromatic aldehydes in the presence of $Co_2(CO)_8$*

Starting material	Reaction product	Temp. (°C)	Yield (%)	Ref.
Benzaldehyde phenyl-hydrazone	N-Phenylphthalimidine	230–240	50	[122]
Benzaldehyde (m-tolyl)-hydrazone	N-(m-Tolyl)phthalimidine	230	20	[122]
1-Naphthaldehyde phenylhydrazone	2-Phenylbenzo(e)-isoindol-1-one	230–235	12	[122]

Experiments to synthesize six membered heterocycles starting from both 1,3-diphenyl-2-propanone phenylhydrazone and 2-ethoxy-1-naphth-aldehyde semicarbazone, which are unable to form five membered rings, failed [692].

The semicarbazones of aromatic ketones also react to form phthalimi-dines [693]. Once again, the expected N-ureido-phthalimidines (shown by the formula in brackets) cannot be isolated, evidently because of hydrogeno-lysis. Moreover, carbonylation probably does not take place exclusively on the unchanged semicarbazone, since benzophenone semicarbazones decompose above their melting points to give benzophenone azines [695]. According to Rosenthal [693] these react with carbon monoxide to give high yields of phthalimidines. The formation of benzhydryl semicarbazones has also been observed; these react further to phenylphthalimides in high yields [693].

4.6. Indazolones and 2,4-Dioxo-1,2,3,4-tetrahydroquinazolines from Azobenzenes

Aromatic azo compounds react with carbon monoxide to form indazolones. At higher temperatures, further reaction of the indazolones leads to insertion of CO to form 2,4-dioxo-1,2,3,4-tetrahydroquinazolines [678, 696].

Thus the reaction of azobenzenes between 170 and 190 °C yields 2-phenylindazolone which reacts with further CO at 220 to 230 °C to form 2,4-dioxo-3-phenyl-1,2,3,4-tetrahydroquinazoline.

Azobenzenes containing o,p-directing substituents react in a similar manner. If only one aromatic ring is substituted, cyclization always takes place on this substituted ring. Azobenzenes containing m-directing substituents do not cyclize [123]. An explanation for this is given in the sections on reaction mechanism.

Table 68. *Reaction of azobenzenes with carbon monoxide in the presence of* $Co_2(CO)_8$

R	Reaction product	Yield (%)	Ref.
H	2-Phenylindazolone	55	[678, 696]
p-CH$_3$	5-Methyl-2-phenylindazolone	35	[678]
p-Cl	5-Chloro-2-phenylindazolone	24	[678]
p-(CH$_3$)$_2$N	5-Dimethylamino-2-phenylindazolone	80	[678]

The reaction of the indazolones by CO insertion may be regarded as analogous to the formation of N,N'-dibenzylurea from hydrazobenzene and carbon monoxide [697].

Table 69. *Products obtained by reaction of azobenzenes with carbon monoxide at temperatures of 200–230 °C in the presence of* $Co_2(CO)_8$

R	R'	Reaction product (2,4-dioxo-1,2,3,4-tetrahydroquinazoline)	Yield (%)	Ref.
H	H	3-Phenyl	65	[678]
p-CH$_3$	H	6-Methyl-3-phenyl	36	[678]
m-CH$_3$	H	7-Methyl-3-phenyl	26	[678]
p-Cl	H	6-Chloro-2-phenyl	45	[678, 696]
p-(CH$_3$)$_2$N	H	6-Dimethylamino-3-phenyl	18	[678]
p-CH$_3$	p-CH$_3$–C$_6$H$_4$	6-Methyl-3-(p-toluyl)-	40	[678]
p-Cl	p-Cl–C$_6$H$_4$	6-Chloro-3-(p-chlorophenyl)	17	[678]
p-CH$_3$O	p-CH$_3$O–C$_6$H$_4$	6-Methoxy-3-(p-methoxy-phenyl)-	28	[678]

The 2,4-dioxo-1,2,3,4-tetrahydroquinazolines can be obtained in a single-stage reaction if the azobenzenes are allowed to react at temperatures above 200 °C [678, 696] (table 69).

4.7. Indones from Cumulenes

Kim and Hagihara [698] studied the carbonylation of tetraphenylbuta-triene in presence of dicobalt octacarbonyl. At 230 to 250 °C and a CO pressure of 150 atm in benzene as solvent they obtained 2-(β,β-diphenyl-vinyl)-3-phenylindone in 70 % yield.

If the reaction is carried out in the presence of water 2-(β,β-diphenyl-ethyl)-3-phenylindanone is formed. This is obviously due to hydrogenation of the two olefinic double bonds [798].

Analogously 1,1-diphenyl-4,4-bis-(4-methoxyphenyl)-butatriene reacted to give 2-(β,β-di-(4-methoxyphenyl)-vinyl)-3-phenyl-indone [798].

Under similar conditions tetraphenyl allene reacted to give 1,1,3-tri-phenylindene in a yield of 41 % o.th. In this reaction carbon monoxide did not take part (1). However, also in this example cyclic carbonyl com-pounds were formed in low yield. 2-Diphenyl methyl-3-phenylindone (2) was obtained in 23 % yield besides 17 % of a product which was assumed to be 2,2,4-triphenyl naphthalinone (3).

These reactions very likely also proceed via the mechanism discussed earlier in this chapter (page 148 ff.).

4.8. Ketones from Dienes

Nonconjugated dienes react in good yield to mixtures of unsaturated and saturated cyclic ketones [699]. These reactions also can be explained by the proposed ring closure mechanism [668]. Good yields are obtained when the position of the two double bonds favors ring closure, which is the case when there are one or two carbons or heteroatoms between the double bonds [700].

Thus, Klemchuk [699] obtained from diallyl a mixture consisting of 2,4-dimethyl-cyclopentene-3-one and 2,5-dimethylcyclopentanone.

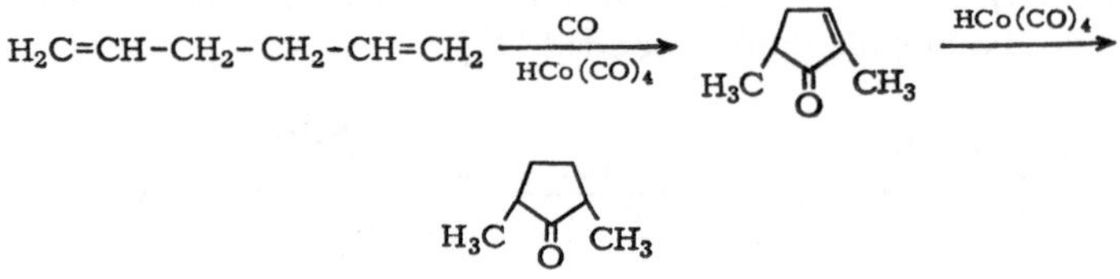

First only one double bond of the diene reacts with cobalt hydrocar-bonyl with formation of π-complexes of the following structures

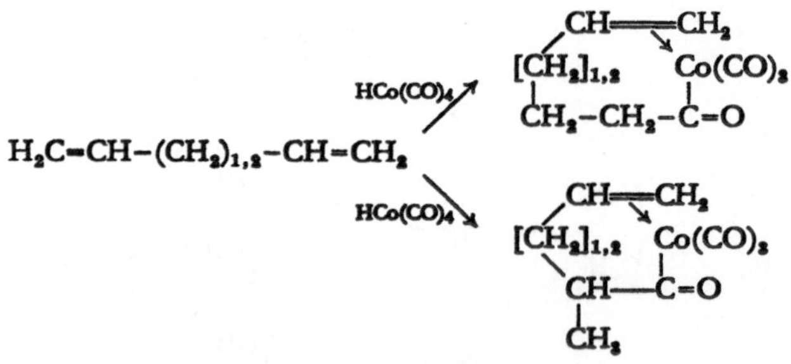

which yield conjugated unsaturated cyclic ketones by ring closure and elimination of cobalt hydrotricarbonyl.

Heck [700] synthesized these π-complexes in another way and observed their reactions to the above-mentioned ketones at temperatures of 25 °C, yields being about 75 %.

The formation of saturated ketones which is frequently observed in this reaction is due to double bond hydrogenation of conjugated unsaturated ketones catalyzed by cobalt hydrocarbonyl. The susceptibility to this type of hydrogenation was already mentioned in the chapter 6.3 on the hydroformylation reaction.

This type of reaction can also be catalyzed by Pd complexes. Thus, e. g. hepta-1,6-diene afforded 5-ethyl-2-methyl-cyclopent-2-enone and 2-ethyl-5-methyl-cyclopent-2-enone in 16 % yield when reacted with CO at 200 °C and 1000 atm in the presence of a 2 % solution of diiodo-bis-(tributylphosphine) palladium (II) [818].

$$H_2C=CH-CH_2-CH_2-CH_2-CH=CH_2 + CO \xrightarrow{PdI_2[(P(C_4H_9)_3]_2}$$

As reported by Heck [700], ring closure can also be effected with compounds in which a heteroatom is placed between the acylium cation and the double bond.

$$H_2C=CH-CH_2-O-CH_2-\overset{\overset{O}{\|}}{C}-Co(CO)_4 \longrightarrow + HCo(CO)_4$$

Analogously cyclic dienes react to give unsaturated bicyclic ketones. Thus, Brewis and Hughes [679] obtained bicyclo-(3.3.1)-non-2-en-9-one in 45 % yield from cyclooctadiene and carbon monoxide in the presence of 1 % diiodo-bis-(tributylphosphine)-palladium(II).

$$\xrightarrow[CO]{HPd(L)_2X} \quad C=O \ + \ HPd(L)_2X$$

The formation of methylcyclohexenone carboxylic acid from methylacetylene, allyl chloride and nickel tetracarbonyl, observed by Chiusoli and Botaccio [680], is also in line with the described mechanism. It can be illustrated by the following formulae.

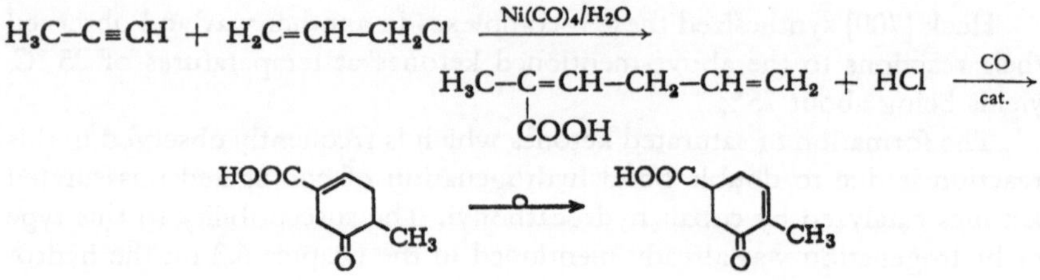

$$H_3C-C\equiv CH + H_2C=CH-CH_2Cl \xrightarrow{Ni(CO)_4/H_2O}$$

$$H_3C-\underset{\underset{COOH}{|}}{C}=CH-CH_2-CH=CH_2 + HCl \xrightarrow[cat.]{CO}$$

4.9. Phenols from Allylhalides, Acetylene and Carbon Monoxide

The above mentioned formation of methylcyclohexanone carboxylic acid, the formation of phenol and the mechanistically unsecured formation of lactones from allyl carbonyl or butyne-1-ol-4 and carbon monoxide are the only known ring closure reactions which are catalyzed by nickel catalysts.

Phenols are also formed as by-products in low yield in the reaction of allyl chloride with acetylene and carbon monoxide in the presence of catalytic amounts of $Ni(CO)_4$.

Higher yields of phenol are obtained [680] if indifferent solvents are used instead of water or alcohols.

The reaction may proceed via the following reaction sequence

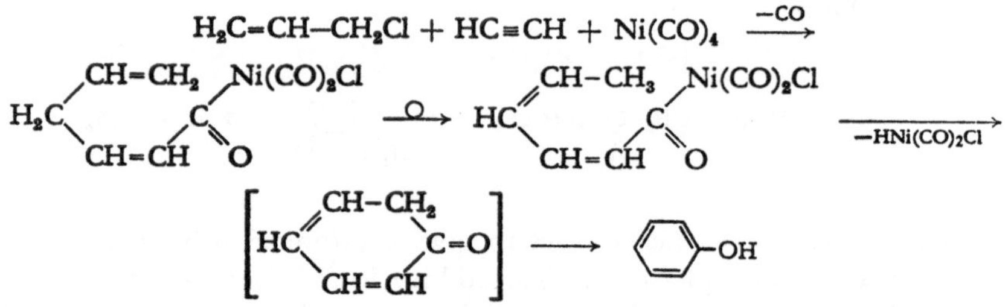

$$H_2C=CH-CH_2Cl + HC\equiv CH + Ni(CO)_4 \xrightarrow{-CO}$$

Analogously, o-cresol is obtained from crotyl chloride and m-cresol from methallyl chloride, whereas phenylacetylene, methallyl chloride and CO yield 2-phenyl-m-cresol. Propyne, allyl chloride and CO react to give o-cresol [680, 799].

4.10. Lactones from Acid Halides, Acetylene and Carbon Monoxide

Lactones are formed from acid halides, acetylene and CO under the same reaction conditions which are applied to allyl halides described in the previous chapter. Ketones may serve as solvents [799, 800].

$$RCOCl + HC{\equiv}CH + Ni(CO)_4 + H_2O \longrightarrow \text{[lactone ring]}_R + Ni(OH)Cl + 3\,CO$$

The cis and trans 2,5-hexadienyl acid chloride isomers react differently:

$$\begin{array}{c} H_2C{=}CH{-}CH_2 \\ \quad\quad\quad\backslash \\ HC{=}CH \\ \quad\quad\backslash \\ COCl \end{array} + HC{\equiv}CH + Ni(CO)_4 + H_2O \longrightarrow H_7C_5\text{[furanone ring]}$$

$$\begin{array}{c} H\quad H \\ \backslash\;/ \\ C{=}C \\ /\quad\backslash \\ H_2C{=}CH{-}H_2C\quad COCl \end{array} + HC{\equiv}CH + Ni(CO)_4 + H_2O \longrightarrow \text{[cyclopentenone]}{-}CH_2{-}\text{[lactone]}$$

$$+ Ni(OH)Cl + 2\,CO$$

In a recent paper Cassar et. al. [801] reported the formation of unsaturated ε-lactones of the type below from acid halides, $Ni(CO)_4$ and acetylene at low concentrations of carbon monoxide.

$$RCOCl + 2\,HC{\equiv}CH + Ni(CO)_4 + H_2O \longrightarrow \text{[ring]}_R{-}O{-}C{=}O$$

$$+ Ni(OH)Cl + 2\,CO$$

4.11. Ring Closure Reactions with Unresolved Reaction Mechanisms

A number of cyclic carbonyl compounds are formed under the conditions of the Reppe reaction or of the Koch acid synthesis. Since the reactions are carried out in the presence of water they can either proceed via

$$HC{\equiv}C{-}CH_2{-}CH_2OH$$

$$\downarrow Ni(CO)_4 \,\big|\, CO/HX$$

$$\begin{array}{c} H_2C{=}C{-}CH_2{-}CH_2OH \\ \quad\quad\;| \\ \quad\quad C{=}O \\ \quad\quad\;| \\ \quad\quad Ni(CO)_2X \end{array} \xrightarrow[-HNi(CO)_2X]{+H_2O} \left[\begin{array}{c} H_2C{=}C{-}CH_2{-}CH_2{-}OH \\ \quad\quad| \\ \quad\quad COOH \end{array} \right] \xrightarrow{-H_2O} \text{[lactone]}\quad (1)$$

$$\left[\begin{array}{c} H_2C\diagdown \\ \quad\quad C{-}CH_2 \\ O{=}C{\to}O{-}CH_2 \\ \quad\oplus\quad\quad| \\ \quad\quad\quad H \\ {[Ni(CO)_2X]}^{\ominus} \end{array} \right] \xrightarrow{-HNi(CO)_2X} \quad (2)$$

the mechanism of the Reppe reaction, or Koch reaction resp., or via the mechanism of the ring closure reaction. The experimental data available so far do not allow a precise decision in favor of one of the discussed mechanisms. Examples are the formation of α-methylene-γ-butyrolactone [429] from butyne-1-ol-4 under the conditions of the Reppe reaction, which can be illustrated as well by equation (1) (mechanism of the Reppe reaction) as by equation (2) (mechanism of the ring closure reaction).

Another example is the carbonylation of allyl carbinol under the conditions of the Reppe reaction [525], yielding α-methyl-γ-butyrolactone plus δ-valerolactone.

The synthesis of lactones from saturated aldehydes under the conditions of the Koch synthesis as described by Himmele [660] could proceed either via the mechanism of the Koch synthesis with dehydration of the hydroxy

$$H_3C\text{—}CH_2\text{—}CH_2\text{—}\underset{\underset{CH_3}{|}}{CH}\text{—}CHO \xrightarrow{\ H^\oplus\ } H_3C\text{—}CH_2\text{—}CH_2\text{—}\underset{\underset{CH_3}{|}}{CH}\text{—}\overset{\oplus}{C}HOH$$

$$\updownarrow$$

$$H_3C\text{—}CH_2\text{—}CH_2\text{—}\underset{\underset{CH_3}{|}}{\overset{\oplus}{C}}\text{—}CH_2OH$$

$$\updownarrow$$

$$H_3C\text{—}CH_2\text{—}\overset{\oplus}{C}H\text{—}\underset{\underset{CH_3}{|}}{CH}\text{—}CH_2OH$$

$$\downarrow CO$$

$$H_3C\text{—}CH_2\text{—}\underset{\underset{\oplus C=O}{|}}{CH}\text{—}\underset{\underset{CH_3}{|}}{CH}\text{—}CH_2OH$$

$$\left[\; H_3C\text{—}CH_2\text{—}\underset{\underset{COOH}{|}}{CH}\text{—}\underset{\underset{CH_3}{|}}{CH}\text{—}CH_2OH \;\right]$$

$$H_3C\text{—}CH_2\text{—}\underset{\underset{\underset{\oplus}{C=O}}{|}}{CH}\text{—}\underset{\underset{CH_3}{|}}{CH}\text{—}CH_2OH$$

$$\left[\begin{array}{c} C_2H_5\text{—}\underset{\underset{O=\overset{\oplus}{C}}{|}}{CH}\text{—}\underset{\underset{CH_2}{|}}{CH}\text{—}CH_3 \\ O \\ Al \\ H \end{array}\right]$$

acid formed as an intermediate or via the mechanism of ring closure reaction with direct attack of the hydroxyl group on the acylium cation.

The same holds for the synthesis of unsaturated γ-lactones starting from unsaturated aldehydes. Thus, the formation of an unsaturated γ-lactone which was reported by Himmele [802] to occur when 2-ethylhexene-1-al was reacted with CO in the presence of 96 % H_2SO_4 at 200 atm and 60 °C in 70.5 % yield, might also follow either the mechanism of the Koch reaction or the ring closure mechanism.

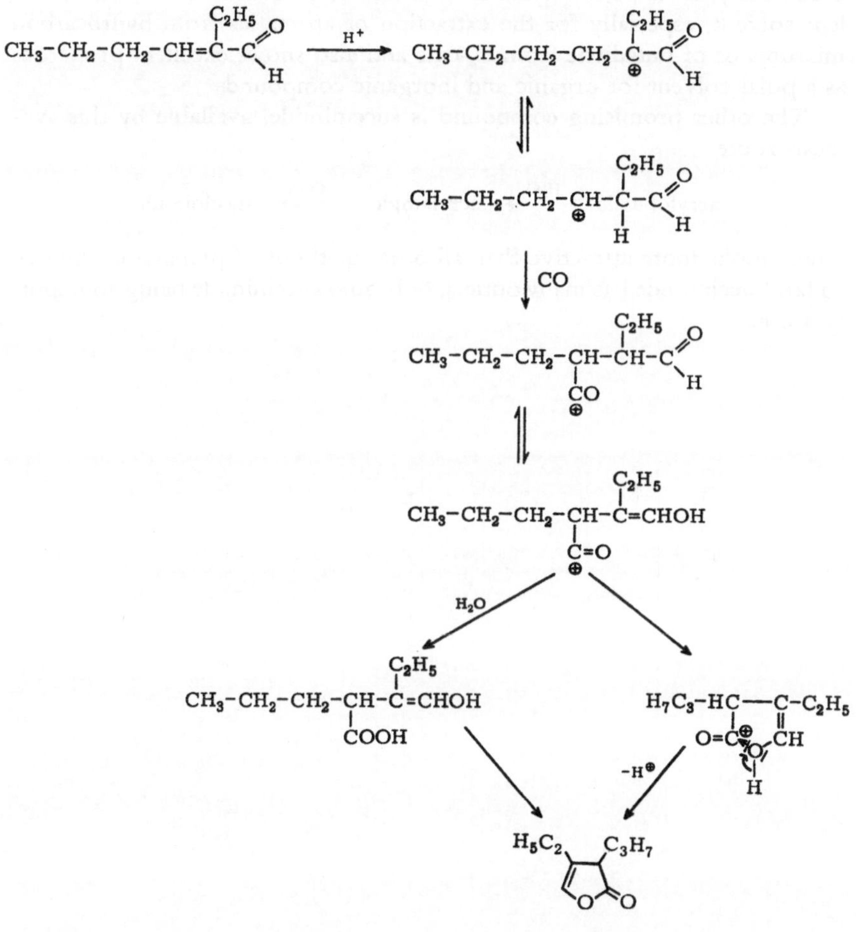

Although the author did not report the exact structure of the C_9-lactone formed, it is very likely that it is an α,β-unsaturated γ-lactone as shown in the above scheme.

5. Technical Aspects

The majority of the ring closure reactions have been reported only recently. Therefore it is not surprising that there is so far no large scale industrial application. However, a number of products obtained are interesting starting materials for pharmaceuticals and pesticides. Thus, α-methylene-γ-butyrolactone is an effective antibiotic [429].

Two of the products described are of greater interest for large-scale production. One is N-methylpyrrolidone (NMP) which is accessible on the basis of allyl chloride, methylamine and carbon monoxide. NMP is an excellent solvent, especially for the extraction of aromatics from hydrocarbon mixtures or of butadiene from C_4-cuts and also shows excellent properties as a polar solvent for organic and inorganic compounds.

The other promising compound is succinimide, available by this synthesis route

$$\text{acrylonitrile} \xrightarrow{\text{H}_2\text{O}} \text{acrylamide} \xrightarrow{\text{CO}} \text{succinimide}$$

which looks more attractive than all other methods of preparation known so far. Succinimide has many outlets, N-bromo succinimide being an important one.

Laboratory Preparations with Carbon Monoxide

Most of the reactions described in this review proceed under pressure. On the laboratory scale stirred or shaken autoclaves are used for discontinous operations.

Carbon monoxide and most of the transition metal carbonyls are highly toxic. For a number of reactions mixtures of carbon monoxide and hydrogen are used. In case of leakages, explosive mixtures of CO/H_2 and air may be generated.

It is therefore recommended that autoclaves be placed in concrete cells or behind concrete walls for safety reasons. The concrete cells should be vented; moreover it is recommended to install instruments for monitoring the carbon monoxide and hydrogen contents in the air of the autoclave cells. A number of publications describe construction and installation of suitable autoclave bunkers [836, 837]. The regulations of the local safety authorities should be carefully observed.

Hydroformylation, ring closure reactions and the Koch acid syntheses may be carried out in autoclaves made from stainless steel. If the formation of iron pentacarbonyl has to be completely avoided, silver or copper lined autoclaves may be used. If hydrogen halides or hydrogen halide generating compounds are used, which e.g., is the case in the Reppe carbonylation reactions, stainless steel autoclaves cannot be used due to corrosion. Hastelloy B or Hastelloy C should be used instead.

Very often it is useful to follow the reaction during the run by taking samples from the autoclaves or the pressure reactors through dip tubes.

For continous operations on the laboratory scale, the set-up is very similar to the technical operations on plant scale as described in the individual chapters. The equipment may be purchased from autoclave manufacturing companies.

Carbon monoxide required for laboratory operations may be purchased in cylinders. If no cylinders are available, carbon monoxide can be prepared by dehydrating formic acid with sulfuric acid or phosphoric acid. The formic acid is added dropwise to the concentrated H_2SO_4 or H_3PO_4 at $120-150C°$. The gas evolved may be compressed with stainless steel compressors. For using carbon monoxide on a very small scale see H. Adkins and G. Krsek, J. Am. Chem. Soc. **70**, 383 (1948) and J. Am. Chem. Soc. **71**, 351 (1949).

For operations with stoichiometric amounts of metal carbonyls see L. Kirch, M. Orchin, J. Am. Chem. Soc. **81**, 3597 (1959) and G. L. Karapinka, M. Orchin, Org. Chemistry **26**, 4187 (1961).

References

1. Hecht, O., Kröper, H.: Naturforschung und Medizin in Deutschland 1939—1946, Vol. 36, Präparative organische Chemie, Part I, p. 115. Edited by K. Ziegler. Wiesbaden: Dieterich'sche Verlagsbuchhandlung 1948.
2. Roelen, O.: Naturforschung und Medizin in Deutschland 1939—1946, Vol. 36, Präparative organische Chemie, Part I, p. 166. Edited by K. Ziegler. Wiesbaden: Dieterich'sche Verlagsbuchhandlung 1948.
3. Kröper, H.: Carbonylierung. In: Ullmanns Encyklopädie der technischen Chemie, Vol. 5, p. 122. München-Berlin: Urban & Schwarzenberg 1954.
4. Pino, P., Paleari, L.: Oxosynthese. In: Ullmanns Encyklopädie der technischen Chemie, Vol. 13, p. 60. München-Berlin: Urban & Schwarzenberg 1962.
5. Schuster, K.: Oxo-Synthese. In: Fortschr. Chem. Forsch. 2, 311—374 (1951).
6. Kröper, H.: Anlagerung von Kohlenmonoxyd und Wasserstoff an Olefine (Hydroformylierung). In: Houben-Weyl, Vol. IV/2, p. 367. Stuttgart: Georg Thieme 1955.
7. Wender, I., Sternberg, H. W., Orchin, M.: Catalysis, Vol. V, p. 73. New York: Reinhold Publishing Corp. 1957.
8. Falbe, J.: Brennstoff-Chem. **45**, 339 (1964).
9. Asinger, F.: Chemie und Technologie der Monoolefine, p. 650. Berlin: Akademie-Verlag 1957.
10. Orchin, M., in: Advances in Catalysis, Vol. V, p. 383. New York: Academic Press 1950.
11. Asinger, F.: Die katalytische Hydrierung des Kohlenoxyds über Kobalt- und Eisenkatalysatoren (Fischer-Tropsch-Synthese). Chem. Techn. der Paraffin-Kohlenwasserstoffe. Berlin: Akademie-Verlag 1956.
12. Pichler, H., in: Advances in Catalysis, Vol. IV, pp. 271, 341. New York: Academic Press 1952.
13. Fell, B., Ulrich, R.: Synthesen mit Kohlenmonoxyd. In: Forschungsberichte des Landes Nordrhein-Westfalen No. 1303. Köln und Opladen: Westdeutscher Verlag 1964.
14. Mond, L.: German Pat. 98643 (1887), Z. 98 II, 1229.
15. Berthelot, M.: Liebigs Ann. Chem. **97**, 125 (1856).
16. Losanitsch, S. M., Jovitschitsch, H. Z.: Ber. **30**, 135 (1897).
17. Schmidt, J.: Das Kohlenmonoxyd. Leipzig: Akad. Verlagsanstalt Goest und Portig KG 1950.
18. Holm, M. M., Reichl, R. H., Vaughan, W. E.: Fiat Report 1000, 1945.
19. Hasche, R. C., Bios (Mt. Vernon, Iowa), 27 (1945).
20. Hall, C. C.: Bios (Mt. Vernon, Iowa), 447 (1945).
21. Roelen, O.: German Pat. 849548 (1938), Z. **1953**, 927; U. S. Pat. 2327066 (1943), C. A. **38**, 550 (1944); Belg. Pat. 436625 (1939), Z. **1941** I, 1354; Fr. Pat. 860289 (1939), Z. **1941** II, 536.
22. — Angew. Chem. A **60**, 213 (1948).

23. — Naturforschung und Medizin in Deutschland, Vol. 36, Präparative orga-
    nische Chemie, Part I, p. 157. Edited by K. Ziegler. Wiesbaden: Dietrich'-
    sche Verlagsbuchhandlung 1948.
24. Adkins, H., Krsek, G. J.: J. Am. Chem. Soc. **71**, 3051 (1949).
25. Keulemans, A. J. M., Kwantes, A., van Bavel, T.: Rec. Trav. Chim. **67**
    298 (1948).
26. Aldridge, G. L., Fasce, E. V., Jonassen, H. B.: J. Phys. Chem. **62**, 869—870
    (1958).
27. — Jonassen, H. B.: Nature **188**, 404 (1960).
28. Macho, V., Mistrik, E. J., Ciha, M.: Collection Czech. Chem. Commun.
    29, 826 (1964).
29. Adkins, H., Krsek, G.: J. Am. Chem. Soc. **70**, 383 (1948).
30. Orchin, M., Kirch, L., Goldfarb, J.: J. Am. Chem. Soc. **78**, 5450 (1956).
31. Karapinka, G., Orchin, M.: Abstracts 137th A. C. S. Meeting, Cleveland,
    Ohio, April 5—14, 1960, p. 92—100.
32. Kirch, L., Orchin, M.: J. Am. Chem. Soc. **80**, 4428 (1958).
33. — Orchin, M., J. Am. Chem. Soc. **81**, 3597 (1959).
34. Wender, I., Sternberg, H. W., Orchin, M.: J. Am. Chem. Soc. **75**, 3041
    (1953).
35. Heck, R. F., Breslow, D. S.: J. Am. Chem. Soc. **83**, 4023 (1961).
36. Breslow, D. S., Heck, R. F.: Chem. Ind. (London) **1960**, 467.
37. Marko, L., Bor, G., Almasy, G., Szabo, P.: Brennstoff-Chem. **44**, 184 (1963).
38. Natta, G., Ercoli, R., Castellano, S.: Chim. Ind. (Milan) **37**, 6 (1955).
39. — — — Barbieri, F. H.: J. Am. Chem. Soc. **76**, 4049 (1954).
40. Martin, A. R.: Chem. Ind. (London) **1954**, 1536.
41. Greenfield, H., Metlin, S., Wender, I.: Abstract of Papers 126th Meeting of
    the American Chemical Soc., New York, Sept. 1954.
42. Chatt, J., Venanzi, L. M.: J. Chem. Soc. **1957**, 4735.
43. Sternberg, H. W., Wender, I.: Chem. Soc. (London) Spec. Publ. **13**, 35
    (1959).
44. Niwa, M., Yamaguchi, M.: Shokubai (Tokyo) **3**, (3), 264—278 (1961).
45. Brenman, A., Herskovits, Z., Herzog, A. M.: Zh. Prikl. Khim. **34**, 454
    (1961).
46. Blanchard, A. A.: Chem. Rev. **21**, 19 (1937).
47. Hieber, W., Fack, E.: Z. Anorg. Allgem. Chem. **236**, 83, 106 (1938).
48. — Hübel, W.: Z. Naturforsch. **7b**, 322 (1952).
49. Reppe, W.: Liebigs Ann. Chem. **582**, 122 (1953).
50. Sternberg, H. W., Wender, I., Friedel, R. A., Orchin, M.: J. Am. Chem.
    Soc. **75**, 2717 (1953).
51. Hieber, W., Hübel, W.: Z. Elektrochem. **57**, 235 (1957).
52. Edgell, W. F., Gallup, G.: J. Am. Chem. Soc. **77**, 5762 (1955).
53. — — J. Am. Chem. Soc. **78**, 4188 (1956).
54. — Magee, C., Gallup, G.: J. Am. Chem. Soc. **78**, 4185 (1956).
55. Cotton, F. A., Wilkonson, G.: Chem. Ind. (London) **1956**, 1305.
56. Bor, G.: Proc. 7th Int. Conf. on Coordination Chem., Stockholm 1962,
    p. 8.
57. — Marko, L.: MAFKI, Ber. d. Ung. Erdöl- und Erdgasforsch. Inst. **3**,
    216 (1962).
58. Takegami, Y., Yokokawa, C., Watanabe, Y., Masada, H., Okuda, Y.: Bull
    Chem. Soc. Japan **37**, 1190 (1964).
59. Falbe, J., Huppes, N., Korte, F. (Shell): German Pat. 1186041 (8. 4. 1961),
    Z. **1965**, 40—2568.

60. — — — Chem. Ber. **97**, 863 (1964).
61. Piacenti, F., Pino, P., Bertolaccini, P. L.: Chim. Ind. (Milan) **44**, 600 (1962).
62. Pino, P., Piacenti, F., Neggiani, P. P.: Chem. Ind. (London) **1961**, 1400.
63. Kniese, W., (BASF): Private communication.
64. Coffield, T. H., Kozikowski, J., Clossen, R. D.: J. Org. Chem. **22**, 598 (1957).
65. Cotton, F. A., Wilkinson, G.: Advanced Inorganic Chemistry, p. 658. New York: Interscience Publishers, Inc. 1962.
66. Calderazzo, F., Cotton, F. A.: Inorg. Chem. **1**, 30 (1962).
67. Coffield, T. H., Closson, H. D., Kozikowski, J.: Abstracts of Conference Papers, Int. Conf. on Coordination Chemistry, London, April 6—11, 1959, Paper No. 26, p. 126.
68. Fischer, E. O., Werner, H.: Angew. Chem. **75**, 57 (1963).
69. Hepner, F. R., Trueblood, K. N., Luens, H. W. J.: J. Am. Chem. Soc. **74**, 1333 (1952).
70. Takegami, Y., Yokokawa, C., Watanabe, Y., Masada, H., Okuda, Y.: Bull. Chem. Soc. Japan **38**, 787 (1965).
71. Brown, H. C., Subbarao, B. C.: J. Org. Chem. **22**, 1137 (1957).
72. — — J. Am. Chem. Soc. **81**, 6434 (1959).
73. — Zweifel, G.: J. Am. Chem. Soc. **82**, 1504 (1960).
74. — Bhatt, M. V.: J. Am. Chem. Soc. **82**, 2074 (1960).
75. — Moerikofer, A W.: J. Am. Chem. Soc. **83**, 3417 (1961).
76. — Hydroboration, p. 140—147. New York: W. A. Benjamin, Inc. Publishers 1962.
77. Speier, J. L., Webster, J. A., Barnes, G. H.: J. Am. Chem. Soc. **79**, 974 (1957).
78. Saam, J. C., Speier, J. L.: J. Am. Chem. Soc. **80**, 4104 (1958).
79. Sellin, T. G., West, R.: J. Am. Chem. Soc. **84**, 1863 (1962).
80. Asinger, F., Fell, B., Janssen, R.: Chem. Ber. **97**, 2515 (1964).
81. — Berg, O.: Chem. Ber. **88**, 445 (1955).
82. Gankin, W. J., Krinkin, D. P., Rudkowskii, D. M.: J. Org. Chem. **2**, 45—46 (1966).
83. Goldfarb, J., Orchin, M., in: Advances in Catalysis, Vol. IX, p. 609. New York: Academic Press 1957.
84. Asinger, F.: Chemie und Technologie der Monoolefine, p. 874. Berlin: Akademie-Verlag 1957.
85. Piacenti, F., Cioni, C., Pino, P.: Chim. Ind. (Milan) **41**, 794 (1959).
86. Karapinka, L., Orchin, M.: J. Org. Chem. **26**, 4187 (1961).
87. Manuel, T. A.: J. Org. Chem. **27**, 3941 (1962).
88. Johnson, M.: J. Chem. Soc. **1963**, 4859.
89. Pino, P., Pucci, S., Piacenti, F.: Chem. Ind. (London) **1963**, 294.
90. Davies, N. R.: Nature **201**, 490 (1964).
91. Harrod, J. F., Chalk, A. J.: J. Am. Chem. Soc. **86**, 1776 (1964).
92. — — Nature **202**, 280 (1965).
93. Rinehart, R. E., Lasky, J. S.: J. Am. Chem. Soc. **86**, 1776 (1964).
94. Falbe, J., Korte, F.: Brennstoff-Chem. **45**, 103 (1964).
95. Höver, H., Mergard, H., Korte, F.: Liebigs Ann. Chem. **685**, 89 (1965).
96. Korte, F., Höver, H.: Tetrahedron **1965**, 1287.
97. Fell, B., Krings, P., Asinger, F.: Chem. Ber. **99**, 3688 (1966).
98. Roos, L., Orchin, M.: J. Am. Chem. Soc. **87**, 5502 (1965).
99. Marko, L.: Chem. Ind. (London) **1962**, 260.
100. Wender, I., Orchin, M., Storch, H. H.: J. Am. Chem. Soc. **72**, 4842 (1960).

101. Pino, P.: Oxosynthese. In: Ullmanns Encyklopädie der technischen Chemie, Vol. 13, p. 61. München-Berlin: Urban & Schwarzenberg 1962.
102. Asinger, F.: Chemie und Technologie der Monoolefine, p. 656. Berlin: Akademie-Verlag 1957.
103. Tramm, H., Kolling,H., Schnur, F., Büchner, K., Heger, H., Stiebling, E. (Ruhrchemie AG): Brit. Pat. 736875 (1955), C. A. **50**, 13982 (1956).
104. Wilson, W., (Standard Oil): U. S. Pat. 2695315 (1954), C. A. **49**, 15945 (1955).
105. Schiller, G. (Chem. Verwertungsges. Oberhausen): German Pat. 953605 (1956), C. A. **53**, 11226 (1959).
106. Esso: Brit. Pat. 801734 (1956), C. A. **53**, 7014 (1959).
107. Büchner, K., (Ruhrchemie AG): German Pat. 874304 (1951), Z. **1954**, 188.
108. Hieber, W.: Angew. Chem. **65**, 534 (1953).
109. Gresham, W. F., McAlvey, A., (Du Pont): U. S. Pat. 2564104 (1951), C. A. **46**, 4561 (1952).
110. Mason, R. B., (Esso): U. S. Pat. 2811567 (1957), C. A. **52**, 4677 (1958).
111. — U. S. Pat. 2754332 (1956), C. A. **51**, 2017 (1957).
112. Aldridge, C. L., (Esso): German Pat. 1125900 (1962), Brit. Pat. 864142 (1961), C. A. **55**, 18597 (1961).
113. Kurokawa, K., Ino, H., Aizawa, R., Amemiya, T.: Nenryo Kyokaishi **41** (422) 539 (1962), C. A. **61**, 11884 (1964).
114. Aldridge, C. L., (Esso): U. S. Pat. 3091644 (1963), C. A. **59**, 11260 (1963).
115. Esso: Brit. Pat. 907027 (1962), C. A. **59**, 3775 (1963).
116. — Brit. Pat. 864142 (1961), C. A. **55**, 18597 (1961).
117. Hieber, W., Schulten, H., Marni, R.: Z. Anorg. Allgem. Chem. **55**, 7, 24 (1942).
118. — — — Z. Anorg. Allgem. Chem. **240**, 261 (1939).
119. Gmelins Handbuch der Anorganischen Chemie, 8th edit., Kobalt Part A, System No. 58, pp. 78 and 346. Weinheim: Verlag Chemie 1961.
120. Iwanaga, R.: Bull. Chem. Soc. Japan **35**, 778 (1962).
121. Heck, R. F, Breslow, D. S.: J. Am. Chem. Soc. **85**, 2779 (1963).
122. Rosenthal, A., Weir, M. R. S.: J. Org. Chem. **28**, 3025 (1963).
123. Murahashi, S., Horiie, S., Jo, T.: Bull. Chem. Soc. Japan **33**, 81 (1960).
124. Hieber, W.: Chemie **55**, 7 (1942).
125. Asinger, F.: Chemie und Technologie der Monoolefine, p. 657. Berlin: Akademie-Verlag 1957.
126. Hughes, V. L., Kirshenbaum, I.: Ind. Eng. Chem. **49**, 1999 (1957).
127. Gemassmer, A. (Chem. Verwertungsges. Oberhausen): German Pat. 884793 (1953).
128. Macho, V.: Acta Chim. **36**, 157/161 (1963).
129. — Acta Chim. **36**, 158 (1963).
130. Hasek, R. H., Wayman, C. E., (Eastman Kodak): U. S. Pat. 2820059 (1958), C. A. **53**, 13040 (1959).
131. Goldfarb, J., Orchin, M.: Advan. Catalysis **9**, 609 (1957).
132. Macho, V., Ciha, M.: Czech. Pat. 103977 (1962), C. A. **60**, 409 (1964).
133. Iwanaga, R.: Bull. Chem. Soc. Japan **35**, 865 (1962).
134. Macho, V.: Acta Chim. **36**, 163 (1963).
135. — Kandidatska dizertacna praca, p. 24—46, Chemicka Fakulta SVST, Bratislava 1961 (Diss. Slow. Techn. Hochschule Bratislava 1961).
136. Macho, V.: Chem. Zvesti **15**, 181 (1961).
137. Marko, L., Bor, G., Klumpp, E.: Chem. Ind. (London) **1961**, 1491.
138. — — Almasy, G.: Chem. Ber. **94**, 847 (1961).

139. — — Klumpp, E., Marko, B., Almasy, G.: Chem. Ber. **96**, 955 (1963).
140. — — — Angew. Chem. **75**, 248 (1963).
141. Klumpp, E., Marko, L., Bor, G.: Chem. Ber. **97**, 926 (1964).
142. Khattab, S. A., Marko, L.: Acta Chim. Acad. Sci. Hung. **45**, 471 (1964).
143. Marko, L., Bor, G.: J. Org. Chem. **30**, 162 (1965).
144. Laky, J., Szabo, P., Marko, L.: Acta Chim. Acad. Sci. Hung. **46**, 247 (1965).
145. Macho, V.: Acta Chim. **36**, 165 (1963).
146. Tetteroo, J. M. J.: Diss. T. H. Aachen 1965.
147. Macho, V., Mistrik, E. J., Ciha, M.: Collection Czech. Chem. Commun. **29**, 826 (1964).
148. Cannell, L. G., Slaugh, L. H., Mullineaux, R. D., (Shell): German Pat. 1186455 (1965), C. A. **62**, 16054 (1965).
149. Eisenmann, J. L., Yamartino, R. L., (Diamond Alkali): Brit. Pat. 941996 (1963), C. A. **59**, 11291 (1963).
150. Taylor, A. C., Ackroyd, N., (I. C. I.): Brit. Pat. 655976 (1951), C. A. **46**, 7584 (1952).
151. Ebel, A., Gemassmer, A., Wenzel, W. (Chem. Verwertungsges. Oberhausen): German Pat. 896341 (1953), C. A. **50**, 3500 (1956).
152. Adams, C. E., Burney, D. E. (Standard Oil): U. S. Pat. 2464916 (1947), C. A. **43**, 5032 (1949).
153. Vlughter, I. C., Keulemans, A. J. M., Hart, M. L. (Shell): U. S. Pat. 2564456 (1951), C. A. **46**, 1582 (1952).
154. Holm, M. M., Nagel, R. H., Reichl, E. H., Vaughan, W. E.: Fiat Report 1000, p. 53.
155. Montecatini: Ital. Pat. 526559 (1955), C. A. **52**, 16656 (1958).
156. Cerveny, W. J. (Standard Oil): U. S. Pat. 2686206 (1954), C. A. **49**, 10999 (1955).
157. Meis, J., Tummes, H. (Ruhrchemie AG): Belg. Pat. 635884 (1963), C. A. **61**, 13195 (1964).
158. Jones, J. K., Hammer, G. P., Fuqua, M. C. (Esso): U. S. Pat. 2757206 (1956), C. A. **51**, 3876 (1957).
159. Krebs, R., Catterall, R. A.: U. S. Pat. 2768974 (1956), C. A. **51**, 9976 (1957).
160. Sneta, H. (Mitsubishi Chem. Ind.): Jap. Pat. 7868 (1954), C. A. **50**, 8705 (1956).
161. Lemke, H.: Supplement mensual a Chim. Ind. (Paris) **89**, No. 4, 118 (1963).
162. — Hydrocarbon Process **1966**, 148.
163. Starr, C. E., Charlet, E. M. (Standard Oil): U. S. Pat. 2636904 (1953), C. A. **48**, 3385 (1954).
164. Asinger, F.: Chemie und Technologie der Monoolefine, p. 688. Berlin: Akademie-Verlag 1957.
165. v. Kutepow, N., Kindler, H., Eisfeld, K., Dettke, K., Jenne, H., Detzer, H. (BASF): German Pat. 1147796 (1960), C. A. **57**, 2076 (1962).
166. — — — — — — German Pat. 1147797 (1960), C. A. **57**, 2076 (1962).
167. Rosenthal, R. W., Schwartzman, L. H., Greco, N. P., Prober, R.: J. Org. Chem. **1963**, 2835.
168. Gresham, W. F., Brooks, R. E.: U. S. Pat. 2497303 (1950), C. A. **44**, 4492 (1950); Brit. Pat. 637999 (1950), C. A. **44**, 9473 (1950).
169. Gwynn, B. H., Hirsch, J. H., (Gulf Research & Dev.): U. S. Pat. 2734922 (1956), C. A. **50**, 16830 (1956).
170. Uchida, H., Todo, N., Ogawa, K.: Rept. Govt. Chem. Ind. Res. Inst. Tokyo **48**, 266 (1953), C. A. **49**, 16266 (1955).

171. Bloch, B. L., Goldwhite, H., Haszeldine, R. N.: J. Chem. Soc. **1966**, 1447.
172. Wakamatsu, H.: Nippon Kagaku Zasshi **85**, (3), 227—231 (1964), C. A. **61**, 13173 (1964).
173. Imyanitov, N. S., Rudkowskii, D. M.: Petrol. Chem. (USSR) **3**, No. 1, 91, (1964), C. A. **60**, 9072 (1964).
174. Falbe, J., Huppes, N., Korte, F.: Brennstoff-Chem. **47**, 207 (1966).
175. — — Brennstoff-Chem. **48**, 24 (1967).
176. — — Belg. Pat. Appl. 33538 (1966).
177. — — Belg. Pat. Appl. 33539 (1966).
178. Osborn, J. A., Wilkinson, G., Young, J. F: Chem. Commun. **2**, 17 (1965).
179. Millidge, A. F., (Distillers): Fr. Pat. 1411602 (1965), C. A. **64**, 598 (1966).
180. Klopfer, O. E., (Ethyl): Brit. Pat. 1111610 (1961).
181. — U. S. Pat. 3050562 (1962), C. A. **57**, 13217 (1963).
182. Gresham, W. F., Brooks, R. E., Bruner, W. H., (Du Pont): U. S. Pat. 2437600 (1948), C. A. **42**, 4196 (1948).
183. Mistrik, E. J., Ciha, M.: Czech. Pat. 106476 (1963), C. A. **60**, 4011 (1964).
184. Haaf, W.: Private communication.
185. Smith, P., Jaeger, H. H., (I. C. I.): Brit. Pat. 966482 (1960), C. A. **61**, 10593 (1964).
186. — — German Pat. 1159926 (1963), C. A. **60**, 14379 (1964).
187. I. C. I.: Austr. Appl. 30352 (1963).
188. Pichler, H., Firnhaber, B., Kioussis, D.: Brennstoff-Chem. **44**, 337 (1964).
189. Evans, D., Osborn, J. A., Jardine, F. H., Wilkinson, G.: Nature, **208**, 1203 (1965).
190. Lonza: Fr. Pat. 1381091 (1963).
191. Uchida, H., Matsuda, A.: Bull. Chem. Soc. Japan **37**, 373 (1964).
192. Whitman, G. M.: U. S. Pat. 2462448 (1946), C. A. **43**, 4287 (1949).
193. Pino, P.: Oxosynthese. In: Ullmanns Encyklopädie der technischen Chemie, Vol. 13, p. 68. München-Berlin: Urban & Schwarzenberg 1962.
194. Wender, I., Sternberg, H. W., Orchin, M.: The Oxo Reaction. In: Catalysis, Vol. V. p. 121. Ed. by P. H. Emmet. New York: Reinhold Publishing Corp. 1955.
195. Asinger, F.: Chemie und Technologie der Monoolefine, p. 658. Berlin: Akademie-Verlag 1957.
196. Piacenti, F., (Montecatini): Belg. Pat. 613606 (1962), C. A. **58**, 451 (1963).
197. — Pino, P., Lazzaroni, R., Branchi, M.: Chem. Soc. **1966**, 488.
198. Brewis, S.: J. Chem. Soc. **1964**, 5014.
199. Wender, I., Metlin, S., Ergun, S., Sternberg, H. W., Greenfield, H.: J. Am. Chem. Soc. **78**, 5401 (1956).
200. Iwanaga, R.: Bull. Chem. Soc. Japan **35**, 869 (1962).
201. Wakamatsu, H., Iwanaga, R., Kato, J.: 12th Annual Meeting of the Chemical Society Japan, Kyoto, April 1959.
202. Barrick, P. L., Pavlic, A. A. (Du Pont): U. S. Pat. 2506571 (1950), C. A. **44**, 7344 (1950).
203. McKeever, C. H., Agnew, G. H. (Rohm & Haas): U. S. Pat. 2533276 (1950), C. A. **45**, 3415 (1951).
204. Habeshaw, J., Thornes, L. S. (Anglo Iranian Comp.): Fr. Pat. 1039669 (1953); Z. **1955**, 5895.
205. Pino, P.: Gazz. Chim. Ital. **81**, 625 (1951).
206. Hagemeyer, H. J., Hull, D. C. (Eastman Kodak): U. S. Pat. 2790832 (1957), C. A. **51**, 15552 (1957).

207. Prichard, W. W. (Du Pont): U. S. Pat. 2517416 (1950), C. A. **45**, 648 (1951).
208. Fritzsche, H., Roelen, O. (Chem. Verwertungsges. Oberhausen): DBP 888097 (1943).
209. Büchner, K., Kühnel, P. (Ruhrchemie AG): German Pat. 837847 (1952). C. A. **51**, 11696 (1957).
210. German Pat. 854216 (1950), Z. **1953**, 4765.
211. Standard Oil Development: German Pat. Appl. St 3513 (1951).
212. Büchner, K. (Chem. Verwertungsges. Oberhausen): Brit. Pat. 719573 (1950), C. A. **50**, 2652 (1956).
213. — (Ruhrchemie AG): Neth. Pat. 76974 (1955), Z. **1958**, 6392.
214. Rudkovskii, D. M., Imyanitov, N. S.: J. Appl. Chem. USSR **35**, 2611 (1962).
215. Iwanaga, R.: Bull. Chem. Soc. Japan **35**, 871 (1962).
216. Guyer, P., Bosshard, E.: Chimia (Aarau) **18**, 131 (1964).
217. Bird, C. W.: Chem. Rev. **62**, 290 (1962).
218. Du Pont: Brit. Pat. 614010 (1948), C. A. **43**, 4685 (1949).
219. Gresham, W. F. (Du Pont): Brit. Pat. 638754 (1950), C. A. **44**, 9473 (1950).
220. — Brooks, R. E., Bruner, W. M. (Du Pont): U. S. Pat. 2549454 (1951), C. A. **44**, 8552 (1951).
221. Hagemeyer, H. J., Hull, D. C. (Eastman Kodak): U. S. Pat. 2694734 (1954), C. A. **49**, 15947 (1955).
222. — — U. S. Pat. 2694735 (1954), C. A. **49**, 15947 (1955).
223. Natta, G., Ercoli, R., Castellano, S. (Montecatini): Ital. Pat. 516716 (1955), C. A. **52**, 1221 (1958).
224. Niwa, A., Kikuchi, Y., Kamimura, S., Onishi, M. (Mitsubishi Chem. Ind.): Jap. Pat. 1107 (1957), C. A. **52**, 4680 (1958).
225. Pino, P., Ercoli, R., Calderazzo, F.: Chim. Ind. (Milan) **37**, 782 (1955).
226. — Paleari, C.: Gazz. Chim. Ital. **81**, 646 (1951).
227. Reppe, W., Friedrich, H., (BASF): German Pat. 897403 (1953), C. A. **50**, 16830 (1956).
228. Staib, J., Guyer, W. R. F., Slotterbeck, O. C. (Esso): U. S. Pat. 2864864 (1958), C. A. **53**, 9063 (1959).
229. Barrich, P. L. (Du Pont): U. S. Pat. 2542747 (1951), C. A. **46**, 7584 (1951).
230. Wender, I., Feldmann, J., Metlin, S., Gwynn, B. H., Orchin, M.: J. Am. Chem. Soc. **77**, 5760 (1955).
231. Adkins, H., Williams, J. L. R.: J. Org. Chem. **17**, 980 (1952).
232. El Daoushy, M. A. F.: Diss., T. H. Aachen 1964.
233. Taylor, A. W. C. (I. C. I.): Brit. Pat. 798541 (1958), C. A. **53**, 2089 (1959).
234. Habeshaw, J., Thornes, L. S. (Anglo-Iranian Oil): Brit. Pat. 702195 (1954), C. A. **49**, 5513 (1955).
235. Taylor, A. W. C., Lamb, S. A. (I. C. I): Brit. Pat. 684673 (1952), C. A. **48**, 1421 (1954).
236. Nienburg, H. J., Gemassmer, A., Eckard, H. (Chem. Verwertungsges. Oberhausen): German Pat. 888687 (1942).
237. Knap, J. E., Cox, N. R., Privette, W. R.: Chem. Eng. Progr. **62**, 4 (1966).
238. Wender, I.: Petrol. Refiner **35**, 197 (1956).
239. Reppe, W., Vetter, H.: Liebigs Ann. Chem. **582**, 133 (1953).
240. v. Kutepow, N., Kindler, H.: Angew. Chem. **72**, 802 (1960).
241. Holm, M. M., Nagel, R. H., Vaughan, E., Reichl, E. H.: Fiat Report 1000, p. 31.
242. Nienburg, H. (BASF): German Pat. 800400 (1950), C. A. **45**, 1625 (1951).

243. Bird, C. W.: Chem. Rev. **62**, 291 (1962).
244. Heck, R. F., Breslow, D. S.: J. Am. Chem. Soc. **83**, 1097 (1961).
245. Husebye, S., Jonassen, H. B.: Acta Chem. Scand. **18**, 1581 (1964).
246. Jonassen, H. B., Stearns, R. J., Kenttämaa, J., Moore, D. W., Whittaker, A. G.: J. Am. Chem. Soc. **80**, 2586 (1958).
247. Prichard, W. W.: Reissue, U. S. Pat. **24**, 653 (1959).
248. Aldrige, C. L., Jonassen, H. B., Pulkkinen, U. E.: Chem. Ind. (London) **1960**, 374.
249. Moore, D. W., Jonassen, H. B., Jogner, T. B.: Chem. Ind. (London) **1960**, 1304.
250. Morikawa, M.: Bull. Chem. Soc. Japan **37**, 379 (1964).
251. Asinger, F.: Chemie und Technologie der Monoolefine, p. 655. Berlin: Akademie-Verlag 1957.
252. Martin, E. V., Busch, H.: Angew. Chem. **74**, 628 (1962).
253. Falbe, J., Huppes, N.: Brennstoff-Chem. **47**, 314 (1966).
254. — — Brennstoff-Chem. **48**, 183 (1967).
255. Büchner, K. (Ruhrchemie AG): Brit. Pat. 765742 (1957), C. A. **51**, 12970 (1957).
256. Inventa AG: Neth. Pat. 298834; Fr. Pat. 1371085 (1964), C. A. **62**, 460 (1965).
257. Natta, G.: Brennstoff-Chem. **36**, 176 (1955).
258. Wilke, G., Pfohl, W., (Studienges. Kohle): German Pat. 1059904 (1959), C. A. **55**, 7321 (1961).
259. Inventa AG: Brit. Pat. 1007627 (1965) C. A. **64**, 4968 (1966).
260. Goetz, R. W., Orchin, M.: J. Am. Chem. Soc. **85**, 2782 (1963).
261. Natta, G., Ercoli, R.: Chim. Ind. (Milan) **34**, 503 (1952).
262. Diels, O., Alder, K.: Liebigs Ann. Chem. **460**, 98 (1928).
263. Alder, K., Stein, G.: Liebigs Ann. Chem. **514**, 198 (1934).
264. Stockhausen, F.: Diss., Univ. Köln 1959.
265. Iwananga, R., Mori, Y., Yoslinde, T. (Ajinomoto Co): Jap. Pat. 8177 (1957), C. A. **52**, 14661 (1958).
266. Pino, P., Paleari, L.: Oxosynthese. In: Ullmanns Encyklopädie der technischen Chemie, Vol. 13, p. 65. München-Berlin: Urban & Schwarzenberg 1962.
267. Hagemeyer, H. J., Hull jr., D. C. (Eastman Kodak): U. S. Pat. 2610203 (1952), C. A. **47**, 5960 (1953).
268. Stolle, M., Bolle, P.: Helv. Chim. Acta **21**, 1551 (1938).
269. Gresham, W. F., Brooks, R. E., Bruner, W. M. (Du Pont): Brit. Pat. 614010 (1948), C. A. **43**, 4685 (1949).
270. — — — U. S. Pat. 2549454 (1951), C. A. **45**, 8552 (1951).
271. — — — U. S. Pat. 2437600 (1948), C. A. **42**, 4196 (1948).
272. — — U. S. Pat. 2402133 (1946), C. A. **40**, 6093 (1946).
273. Unpublished work BASF, cited in: Houben-Weyl, Vol. VII, 1, p. 61. Stuttgart: Georg Thieme 1954.
274. Kato, J., Wakamatsu, H., Komatsu, T.: Kogyo Kagaku Zasshi **64**, 2139 (1961), C. A. **57**, 2064 (1962).
275. — — Ishihara, H. (Asinomoto & Co): Jap. Pat. 2574 (1961), C. A. **56**, 9977 (1962); Brit. Pat. 838737 (1960), C. A. **55**, 384 (1961); U. S. Pat. 2978481 (1961), C. A. **55**, 15351 (1961).
276. Kodoma, Sh., Taniguchi, J., Takekami, Y.: Jap. Pat. 7770 (1957), C. A. **52**, 13777 (1958).
277. Noguchi Research Foundation: Fr. Pat. 1370004 (1964), C. A. **63**, 9823 (1965).

278. Schreyer, R. C. (Du Pont): U. S. Pat. 2564131 (1951), C. A. **46**, 9583 (1952).
279. Orchin, M., Wender, I.: Catalysis, Vol. I, p. 1. New York: Reinhold Publishing Corp. 1957.
280. Falbe, J., Schulze-Steinen, H. J., Korte, F.: Chem. Ber. **98**, 886 (1965).
281. Hull, D. C., Pery, M. A., Hagemeyer, jr., H. J. (Eastman Kodak): Fr. Pat. 1400958 (1965).
282. Gresham, W. F., Brooks, R. E. (Du Pont): U. S. Pat. 2497303 (1950), C. A. **44**, 4492 (1950).
283. Falbe, J., Korte, F.: Chem. Ber. **97**, 1104 (1964).
284. Rosenthal, A., Abson, D.: Can. J. Chem. **42**, 1811 (1964).
285. — Koch, J.: Can. J. Chem. **42**, 2025 (1964).
286. — Abson, D.: J. Am. Chem. Soc. **86**, 5396 (1964).
287. Habeshaw, J., Geach, C. J. (Anglo-Iranian Oil): Brit. Pat. 702206 (1954), C. A. **49**, 5514 (1955).
288. — Rae, R. W. (Anglo-Iranian Oil): Brit. Pat. 702201 (1954), C. A. **49**, 5514 (1955).
289. Marko, H., Stresinka, J., Macho, V., Gregor, F.: Czech. Pat. 110106 (1964), C. A. **61**, 3025 (1964).
290. Rudkowskii, D. M., Imyanitov, N. S., Gankin, V. Y.: C. A. **57**, 10989 (1962).
291. Gut, G., El Makhzangi, M. H., Guyer, A.: Helv. Chim. Acta **48**, 1151 (1965).
292. Lonza: Neth. Pat. Appl. 6406299 (1964).
293. Greenfield, H., Wotiz, J. H., Wender, I.: J. Org. Chem. **22**, 542 (1957).
294. Jardine, F. H., Osborn, J. A., Wilkinson, G., Young, J. F.: Chem. Ind. (London) **1965**, 560.
295. Orchin, M., in: Advances in Catalysis, Vol. V, p. 401. New York: Academic Press 1953.
296. Heck, R. F., Breslow, D. S.: Chem. Ind. (London) **1960**, 467.
297. Takegami, Y., Yokokawa, C., Watanabe, Y., Masada, H.: Bull. Chem. Soc. Japan **37**, 672 (1964).
298. Yokokawa, C., Watanabe, Y., Takegami, Y.: Bull. Chem. Soc. Japan **37**, 677 (1964).
299. Niederhauser, W. D., (Rohm & Haas): U. S. Pat. 3054813 (1962), C. A. **58**, 3319 (1963).
300. — German Pat. 1252497 (1963), C. A. **50**, 3500 (1956).
301. Heck, R. F.: J. Am. Chem. Soc. **85**, 1460 (1963).
302. Yokokawa, C., Watanabe, Y., Takegami, Y.: Bull. Chem. Soc. Japan **37**, 678 (1964).
303. Takegami, Y., Yokokawa, C., Watanabe, Y.: Bull. Chem. Soc. Japan **38**, 787 (1965).
304. — — — Masada, H.: Bull. Chem. Soc. Japan **38**, 1649 (1965).
305. Roos, L., Goetz, R. E., Orchin, M.: J. Org. Chem. **30**, 3023 (1965).
306. Witzel, G., Scheuermann, A., Kotschmar, A., Eder, K., (BASF): German Pat. 843876 (1941).
307. — Eder, K., Scheuermann, A. (BASF): German Pat. 867849 (1941).
308. Wender, I., Levine, R., Orchin, M.: J. Am. Chem. Soc. **71**, 4160 (1949).
309. Pieroll, K. (BASF): German Pat. 875802 (1941), Z. **1953**, 8201.
310. Müller-Cunradi, M., Pieroli, K., Lorenz, L., Beckmann, H. (BASF): German Pat. 890945 (1942), Z. **1954**, 2048.
311. Bird, C. W.: Chem. Rev. **62**, 283 (1962).

312. Brooks, R. E. (Du Pont): U. S. Pat. 2457204 (1948), C. A. **43**, 3443 (1949).
313. Wender, I., Friedel, R. A., Orchin, M.: Science **113**, 206 (1951).
314. Ziesecke, K. H.: Brennstoff-Chem. **33**, 385 (1952).
315. Kröper, H., Hauber, H., Hagen, W. (BASF): German Pat. 921936 (1955), C. A. **53**, 222 (1959).
316. Monkemeyer, K. (Chem. Werke Hüls): U. S. Pat. 2770655 (1956), C. A. **51**, 5817 (1957).
317. Wender, I., Metlin, S., Orchin, M.: J. Am. Chem. Soc. **73**, 5704 (1951).
318. — Greenfield, H., Metlin, S., Orchin, M.: J. Am. Chem. Soc. **74**, 4079 (1952).
319. Uchida, H., Matsuda, A.: Bull. Chem. Soc. Japan **36**, 1351 (1963).
320. Bertrand, J. A., Aldridge, C. L., Husebeye, S., Jonassen, H. B.: J. Org. Chem. **29**, 790 (1964).
321. Naragon, E. A., Millendorf, A. J., Vergilio, J. H. (Texas Co.): U. S. Pat. 2699453 (1955), C. A. **50**, 1893 (1956).
322. Natta, G., Pino, P., Ercoli, R.: J. Am. Chem. Soc. **74**, 4496 (1952).
323. Wender, I., Levine, R., Orchin, M.: J. Am. Chem. Soc. **72**, 4375 (1950).
324. Berty, I., Marko, L.: Acta Chim. **1952**, 177.
325. Marko, L.: Proc. Chem. Soc. **1962**, 67.
326. Falbe, J., Huppes, N., Korte, F., (Shell): German Pat. Appl. S 96135 (1965).
327. — — German Pat. Appl. S 98107 (1965).
328. Heil, B., Marko, L.: Chem. Ber. **99**, 1086 (1966).
329. Falbe, J.: Unpublished results.
330. Nienburg, H. J., Gemassmer, A. (Chem. Verwertungsges. Oberhausen): German Pat. 902491 (1954), C. A. **50**, 3501 (1956).
331. Parker, P. T., Hilland, G. O. (Standard Oil): U. S. Pat. 2571160 (1951), C. A. **46**, 3555 (1952).
332. Marko, L., Szabo, P.: Ung. Mineralöl- und Erdgas-Versuchsanstalt, p. 296, Publ. 1963.
333. Schuster, C., Eilbracht, H. (Chem. Verwertungsges. Oberhausen): German Pat. 888842 (1942), publ. 1953, Z. **1955**, 3731.
334. Habeshaw, J., Geach, C. J. (Anglo-Iranian Oil): Brit. Pat. 702241 (1948), C. A. **50**, 5515 (1955).
335. Taylor, A. W. C. (I. C. I.): Brit. Pat. 740541 (1955), C. A. **50**, 1683 (1956).
336. Büchner, K., Kühnel, P. (Ruhrchemie AG): German Pat. 879837 (1953), C. A. **50**, 5019 (1956).
337. Koontz, J. D. (Standard Oil): U. S. Pat. 2679534 (1954), C. A. **48**, 10329 (1954).
338. Catterall, W. E. (Esso): U. S. Pat. 2779974 (1957), C. A. **51**, 9674 (1957).
339. I. C. I.: Fr. Pat. 1277098 (1960).
340. Hohenschutz, H.: European Chemical News, Large Plant Supplement, p. 7, Sept. 1966.
341. Hatch, L. F.: Higher Oxo Alcohols. New York: John Wiley 1957.
342. European Chemical News **1962**, 25.
343. Nachrichten aus Chemie und Technik **11** (9), 160 (1963).
344. Chem. Ind. Düsseldorf **1963**, 777.
345. Reppe, W.: Liebigs Ann. Chem. **582**, 1 (1953).
346. — Kröper, H.: Liebigs Ann. Chem. **582**, 38 (1953).
347. — — v. Kutepow, N., Pistor, H. J.: Liebigs Ann. Chem. **582**, 78 (1953).
348. — — Pistor, H. J., Weissbarth, O.: Liebigs Ann. Chem. **582**, 87 (1953).

349. — Neue Entwicklungen auf dem Gebiet der Chemie des Acetylens und Kohlenoxyds, p. 96—126. Berlin-Göttingen-Heidelberg: Springer 1949.
350. Ziegler, K.: Naturforschung und Medizin in Deutschland (1939—1946), Vol. 36, Präparative organische Chemie, Part. I. Wiesbaden: Dieterich'sche Verlagsbuchhandlung 1948.
351. Hecht, O., Kröper, H.: Naturforschung und Medizin in Deutschland (1939—1946), Vol. 36, Präparative organische Chemie, Part I, Sect. 1, p. 115—154. Wiesbaden: Dieterich'sche Verlagsbuchhandlung 1948.
352. Kröper, H., in: Houben-Weyl, Vol. IV/2, p. 385—415. Stuttgart: Georg Thieme 1955.
353. Bird, C. W.: Chem. Rev. 62, 283 (1962).
354. Eidus, Ya. T., Puzitskii, K. V.: Russ. Chem. Rev. 33, 438 (1964).
355. Yamamoto, K., Sato, K. (Mitsui Chem. Ind.): Jap. Pat. 212/53, C. A. 48, 308 (1954).
356. Bird, C. W., Hudec, J.: Chem. Ind. (London) 1959, 570.
357. Almasy, M., Szabo, L., Farkas, J., Bota, T.: Acad. Rep. Populare Romine, Studii Cercetari Chim. 8, 495 (1960), C. A. 55, 19427 (1961).
358. Davison, A., McFartane, N., Pratt, L.: J. Chem. Soc. 1962, 3652.
359. Kröper, H., in: Houben-Weyl Vol. IV/2, p. 387. Stuttgart: Georg Thieme 1955.
360. Krüerke, U., Hübel, W.: Chem. Ind. (London) 1960, 1264.
361. Heck, R. F.: J. Am. Chem. Soc. 85, 2013 (1963).
362. Wender, I., Sternberg, H. W., Orchin, M.: Catalysis, Vol. V, p. 98. New York: Reinhold Publishing Corp. 1957.
363. Behrens, H., Lohöfer, F.: Chem. Ber. 94, 1391 (1961).
364. — Zielsparger, H.: J. prakt. Chem. 14, 249 (1961).
365. Ehrreich, J. E., Nickerson, R. G., Ziegler, C. E.: Ind. Eng. Chem. Process Design Develop. 4, 77 (1965).
366. Thomson, H. W.: J. Chem. Soc. 1943, 522, 1822.
367. Jones, E. R. H., Shen, T. Y., Whiting, M. C.: J. Chem. Soc. 1951, 766.
368. Tetteroo, J. M. J.: Diss., T. H. Aachen (1965), p. 67.
369. Jones, E. R. H., Shen, T. Y., Whiting, M. C.: J. Chem. Soc. 1951, 763.
370. Reppe, W., v. Kutepow, N., Titzenthaler, E. (BASF): German Pat. 922231 (1952), C. A. 51, 16517 (1957).
371. v. Kutepow, N., Himmele, W., Hohenschutz, H.: Chem. Ingr.-Techn. 37, 383 (1965).
372. Kröper, H., in: Houben-Weyl, Vol. IV/2, p. 401, Stuttgart: Georg Thieme 1955.
373. Himmele, W., v. Kutepow, N. (BASF): German Pat. 1104938 (1959).
374. Kröper, H.: Carbonylierung. In: Ullmanns Encyklopädie der technischen Chemie, Vol. 5, p. 123. München-Berlin: Urban & Schwarzenberg 1954.
375. — Ausführung heterogener katalytischer Reaktionen II. In: Houben-Weyl, Vol. IV/2, p. 400. Stuttgart: Georg Thieme 1955.
376. Noble, M. L., (I. C. I.): Brit. Pat. 713325 (1954). C. A. 50, 6500 (1956).
377. Reppe, W., Kröper, H., Pistor, H. J. (BASF): German Pat. 763693 (1941).
378. — Liebigs Ann. Chem. 582, 1 (1953).
379. BASF: Brit. Pat. 802544 (1958), C. A. 53, 7017 (1959).
380. Hagemeyer, jr., J. (Eastman Kodak): U. S. Pat. 2593440 (1952).
381. Samow, G.: Diss., T. H. Aachen 1963.
382. Stadler, R., Henkel, E., Raber, E.: Acrylsäure und Acrylester. In: Festschr Carl Wurster, 60. Geburtstag, p. 65—70, BASF. Ed. J. Weisbecker, Frankfurt 1960.

383. Reppe, W., Reicheneder, F., Stengel, G., Ziegler, A. (BASF): German Pat. 1076672 (1956), Z. **1961**, 3150.
384. — German Pat. 888099 (1950), Z. **1955**, 9194.
385. Friedrich, H., Hoffmann, H. (BASF): German Pat. 1042572 (1958), Z. **1960**, 2005.
386. Reppe, W., Magin, A. (BASF): German Pat. 1173458 (1961), Z. **1965**, 17—2432.
387. — — German Pat. 1215139 (1963).
388. Natta, G., Pino, P.: Chim. Ind. (Milan) **31**, 245 (1949).
389. Pino, P., Miglierina, A.: J. Am. Chem. Soc. **74**, 5551 (1952).
390. Natta, G.: Chim. Ind. (Milan) **34**, 449 (1952).
391. Pino, P., Pietra, E., Mondello, B.: Gazz. Chim. Ital. **84**, 453 (1954).
392. Sternberg, H. W., Shukys, I. G., Donne, C. D., Markby, R., Friedel, R. A., Wender, I.: J. Am. Chem. Soc. **81**, 2339 (1959).
393. Natta, G., Pino, P. (Lonza): U. S. Pat. 2851486 (1958), C. A. **53**, 3153 (1959).
394. — — Swiss. Pat. 291730 (1953), C. A. **49**, 1789 (1955).
395. Lonza: Brit. Pat. 733123 (1955), C. A. **50**, 8717 (1956).
396. — Fr. Pat. 1148924 (1957), Z. **1959**, 3643.
397. Reppe, W., Alberts, H., Friedrich, H. H. (BASF): German Pat. 1040526 (1958), Z. **1959**, 4966.
398. Pino, P., Miglierina, A., Pietra, E.: Gazz. Chim. Ital. **84**, 443 (1954).
399. — — — Gazz. Chim. Ital. **84**, 443 (1954).
400. Reppe, W., Magin, A. (BASF): German Pat. 870698 (1944), Z. **1955**, 8032.
401. Kröper, H., in: Houben-Weyl, Vol. IV/2, p. 665. Stuttgart: Georg Thieme 1955.
402. Lonza: Fr. Pat. 1430131 (1964).
403. Reppe, W., Magin, A. (BASF): German Pat. 849693 (1948), Z. **1953**, 1083.
404. — — German Pat. 880297 (1948), Z. **1955**, 3486.
405. — — German Pat. 894558 (1948), Z. **1954**, 9390.
406. v. Kutepow, N., Bittler, K., Neubauer, D. (BASF): German Pat. 1227023 (1966).
407. — Kindler, H.: Angew. Chem. Intern. Ed. Engl. **1961**, 41.
408. Reppe, W., Schweckendiek (BASF): German Pat. 805641 (1948), Z. **1951**, 2251.
409. — German Pat. 892445 (1951), Z. **1954**, 5173.
410. — Kröper, H., v. Kutepow, N., Pistor, H. J.: Liebigs Ann. Chem. **582**, 79 (1953).
411. Agnst, R.: Diss., T. H. Zürich, Prom. Nr. 2336.
412. Miziroki, T., Nakayama, M., Ando, Y., Furumi, M.: Kogyo Kagaku Zasshi **65**, 1049 1962.
413. — — Furumi, M.: Kogyo Kagaku Zasshi **65**, 1054 1962.
414. Berty, J., Marko, L., Kallo, D.: Chem. Tech. (Berlin) **8**, 260 (1956).
415. Natta, G., Ercoli, R., Castellano, S., Barbieri, F. H.: J. Am. Chem. Soc. **76**, 4049 (1954).
416. Rosenthal, R. W., Schwartzman, L. H.: J. Org. Chem. **24**, 836 (1959).
417. Bird, C. W., Hudec, J.: Chem. Ind. (London) **1960**, 1264.
418. Bauld, N. L.: Tetrahedron Letters **1963**, 1841.
419. Reppe, W., v. Kutepow, N.: German Pat. 893499 (1950), Z. **1954**, 3106.
420. — — Titzenthaler, E. (BASF): German Pat. 899798 (1951), Z. **1954**, 8444.
421. Ohashi, K., Suzuki, S., Ito, H.: J. Chem. Soc. Japan, Ind. Chem. Sect. **55**, 120 (1952).

422. — — — J. Chem. Soc. Japan, Ind. Chem. Sect. **55**, 607 (1952), C. A. **48**, 5793 (1954).
423. Suzuki, S., Uno, K., Yonezawa, M., Ito, H.: J. Chem. Soc. Japan, Ind. Chem. Sect. **55**, 718 (1952), C. A. **49**, 3006 (1955).
424. Caldwell, J. R., (Eastman Kodak): U. S. Pat. 2739158 (1956), C. A. **50**, 15579 (1956).
425. Jones, E. R. H., Shen, T. Y., Whiting, M. C.: J. Chem. Soc. **1951**, 48.
426. Kröper, H.: Carbonylierung. In: Houben-Weyl, Vol. IV/2, p. 414. Stuttgart: Georg Thieme 1955.
427. Du Pont, G., Pignaniol, P., Vialle, J.: Bull. Soc. Chim. France **1948**, 529.
428. Yakubovich, A. Y., Volkova, E. V.: Dokl. Akad. Nauk SSSR **84**, 1183 (1952).
429. Jones, E. R. H., Shen, T. Y., Whiting, M. C.: J. Chem. Soc. **1950**, 230.
430. Mueller, G. P., MacArthur, F. C.: J. Am. Chem. Soc. **76**, 4621 (1954).
431. Jones, E. R. H., Whitham, G. H., Whiting, M. C.: J. Chem. Soc. **1954**, 1865.
432. Shell: Neth. Pat. Appl. 294714 (1965), C. A. **63**, 13442 (1965).
433. Bergmann, E. D., Zimkin, E.: J. Chem. Soc. **1950**, 3455.
434. Chiusoli, G. P.: Gazz. Chim. Ital. **89**, 1331 (1959).
435. Ashworth, P. J., Whitham, G. H., Whiting, M. C.: J. Chem. Soc. **1957**, 4633.
436. Jones, E. R. H., Whitham, G. H., Whiting, M. C.: J. Chem. Soc. **1957**, 4628.
437. Schwartzman, L. H., Rosenthal, R. W.: Abs. A. C. S. Meeting, April 1959, p. 57.
438. Chiusoli, G. P.: Angew. Chem. **72**, 74 (1960).
439. — Chim. Ind. (Milan) **43**, 259, 638 (1961).
440. Sternberg, H. W., Markby, R., Wender, I.: J. Am. Chem. Soc. **82**, 3638 (1960).
441. — Shukys, I. G., Donne, C. D., Markby, R., Friedel, R. A., Wender, I.: J. Am. Chem. Soc. **81**, 2339 (1959).
442. Chiusoli, G. P., Merzoni, S., Mondelli, G.: Tetrahedron Letters **1964**, 2777.
443. — Cassar, L.: Angew. Chem. **79**, 177—186 (1967).
444. Howk, B. W., Sauer, J. C. (Du Pont): German Pat. 1135486, Z. **1963**, 19023.
445. Reppe, W., (BASF): German Pat. 874910 (1951), Z. **1953**, 8459.
446. Tsuji, J., Nogi, T.: J. Org. Chem. **31**, 2641 (1966).
447. Jacobsen, G., Späthe, H. (Farbwerke Hoechst): German Pat. 1138760 (1960), Z. **1964**, 34—2395.
448. v. Kutepow, N., Bittler, K., Neubauer, D. (BASF): German Pat. 1221224 (1966).
449. Yamamoto, K., Sato, K. (Mitsui Chem. Ind.): Jap. Pat. 3763, (1956), C. A. **51**, 14788 (1957).
450. Maki, M.: J. Fuel Soc. Japan **32**, 410 (1953), C. A. **49**, 851 (1955).
451. Smolin, E. M. (American Cyanamid): U. S. Pat. 3025319 (1962), C. A. **57**, 11027 (1962).
452. Reppe, W.: Liebigs Ann. Chem. **582**, 26 (1953).
453. Dunn, J. T. (Union Carbide): Brit. Pat. 879009 (1961), Z. **1964**, 9—2220.
454. — (Union Carbide), U. S. Pat. 3019256 (1961), C. A. **56**, 12747 (1962).
455. Reppe, W.: Liebigs Ann. Chem. **582**, 25 (1953).
456. National Lead Co: Brit. Pat. 887433 (1962), C. A. **57**, 11027, (1962).
457. Reppe, W.: Liebigs Ann. Chem. **582**, 16 (1953).

458. Tsuji, J., Nogi, T.: J. Am. Chem. Soc. **88**, 1289 (1966).
459. Sauer, J. C. (Du Pont): U. S. Pat. 3097237 (1963).
460. v. Kutepow, N., Bittler, K., Neubauer, D., Reis, H. (BASF): German Pat. Appl. Pend.
461. Raczynski, W. A. (Hercules Powder): U. S. Pat. 2738368 (1956), C. A. **50**, 15577 (1956).
462. Reppe, W.: Liebigs Ann. Chem. **582**, 34 (1953).
463. v. Kutepow, N., Bittler, K., Neubauer, D., Reis, H., (BASF): German Pat. Appl. Pend.
464. Tsuji, J., Morikawa, M., Iwamoto, N.: J. Am. Chem. Soc. **86**, 2095 (1964).
465. Reppe, W.: Liebigs Ann. Chem. **582**, 13 (1953).
466. — Hecht, O., Gassenmeier, E. (BASF): German Pat. 851339 (1939), Z. **1954**, 1311.
467. — — — German Pat. 857634 (1940).
468. — — — German Pat. 859611 (1940).
469. v. Kutepow, N., Bittler, K., Neubauer, D. (BASF): German Pat. 1229089 (1966).
470. — — — Belg. Pat. 674069 (1966).
471. — — Meier, F., Neubauer, D.: Belg. Pat. 679611 (1966).
472. Wiedemann, O. (BASF): German Pat. 848357 (1952), Z. **1953**, 130.
473. McKoy, J. W., Swanson, N. (Dow Chemical): U. S. Pat. 3151155 (1964), C. A. **62**, 448 (1965).
474. Tetteroo, J. M. J.: Diss. T. H. Aachen 1965, p. 70.
475. Kröper, H.: Carbonylierung. In: Ullmanns Encyklopädie der technischen Chemie, Vol. 5, p. 129. München-Berlin: Urban & Schwarzenberg 1954.
476. — Anlagerung von Kohlenmonoxyd und Wasserstoff an Olefine (Hydroformylierung). In: Houben-Weyl, Vol. IV/2, p. 387. Stuttgart: Georg Thieme 1955.
477. Levering, D. R., Glaserbrook, A. L.: J. Org. Chem. **23**, 1836 (1958).
478. Gresham, W. R., Brooks, R. E. (Du Pont): U. S. Pat. 2448368, C. A. **43**, 669 (1949).
479. — — Brit. Pat. 631001 (1950), C. A. **44**, 4493 (1950); U. S. Pat. 2549453 (1951), C. A. **45**, 8551 (1951).
480. Hagemeyer, H. J.: U. S. Pat. 2739169 (1952), C. A. **50**, 16, 835 (1956).
481. Larson, A. T.: U. S. Pat. 2448375 (1945), C. A. **43**, 670 (1949).
482. Reppe, W., Kröper, H.: German Pat. 863194 (1943), C. A. **48**, 1425 (1954).
483. — — Liebigs Ann. Chem. **582**, 38 (1953).
484. Ercoli, R.: Chim. Ind. (Milan) **37**, 1029 (1955).
485. Dupont, G., Pignaniol, P., Vialle, J.: Bull. Soc. Chim. France **1948**, 529.
486. Rosenthal, R. W. (Texas Co.): U. S. Pat. 2652413 (1953), C. A. **48**, 5209 (1954).
487. Ercoli, R., Signorini, G., Santabrogio, E.: Chim. Ind. (Milan) **42**, 587 (1960).
488. Reppe, W., Schlichting, O., Klager, K., Toepel, T.: Liebigs Ann. Chem. **560**, 1 (1948).
489. Bird, C. W., Cookson, R. C., Hudec, J.: J. Chem. and Ind. 20, (1960) and unpublished results.
490. Bhattacharyya, S. K., Sourirajan, J.: J. Sci. Ind. Res. (India) **11**B, 123 (1952), C. A. **47**, 10475 (1953).
491. Detzer, H., Metzger, H., Urbach, H. (BASF): Belg. Pat. 613730 (1961), C. A. **58**, 456 (1963).
492. Rull, T.: Mémoires présentes à la Société Chimique No 440, 2680 (1964).

493. Reppe, W., Kröper, H.: Liebigs Ann. Chem. **582**, 60 (1953).
494. Imyanitov, N. S., Kuraev, B. L., Rudkovskii, D. M.: Zh. Prikl. Khim. **38** (11), 2558 (1965), C. A. **64**, 6484 (1966).
495. — Rudkovskii, D. M.: Zh. Organ. Khim. **2** (2) 231 (1966).
496. Esso: U. S. Pat. 3253081 (1962).
497. Jap. Pat. Publ. Nr. 6575/1966 (1966), Rule No. 11 788/1963.
498. Kealy, T. J., Benson, R. E.: J. Org. Chem. **26**, 3126 (1961).
499. Reppe, W., Kröper, H. (BASF): German Pat. 861243 (1943), Z. **1954**, 2493.
500. Büchner, K., Roelen, O., Meis, J., Langwald, H. (Ruhrchemie AG): German. Pat. 1109661 (1959), Z. **1962**, 2903.
501. Reppe, W., Kröper, H. (BASF): German Pat. 868149 (1942), Z. **1953**, 5255.
502. — v. Kutepow, N., Detzer, H. (BASF): German Pat. 1006849 (1956).
503. — (BASF): German Pat. 888099 (1950), Z. **1955**, 9194.
504. — Kröper, H.: Liebigs Ann. Chem. **582**, 44 (1953).
505. Pino, P., Ercoli, R.: Chim. Ind. (Milan) **36**, 536 (1954).
506. Matsuda, A., Uchida, H.: Bull. Chem. Soc. Japan **38**, 710 (1965).
507. Kemchah, P. P. (Esso): U. S. Pat. 3064040 (1962), C. A. **58**, 11221 (1963).
508. Tsuji, J., Hosaka, S., Kiji, J., Susuki, T.: Bull. Chem. Soc. Japan **39**, 141 (1966).
509. — Nogi, T.: Bull. Chem. Soc. Japan **39**, 146 (1966).
510. Montecatini: Brit. Pat. 981008 (1965), C. A. **62**, 7895 (1965).
511. Tsuji, J., Hoskaka, S.: J. Chem. Soc. Japan **1965**, 4075.
512. Jenner, E. L., Lindsey, jr., R. V. (Du Pont), U. S. Pat. 2876254 (1959), C. A. **53**, 17906 (1959).
513. Tsuji, J., Morikawa, M., Kiji, J.: Tetrahedron Letters **1963**, 16, 1061.
514. — — — J. Am. Chem. Soc. **86**, 4851 (1964).
515. Blackham, A. U. (National Distillers): U. S. Pat. 3119861 (1964), C. A. **60**, 14631 (1964).
516. Tsuji, J., Morikawa, M., Kiji, J.: Tetrahedron Letters No. 13, 817 (1965).
517. v. Kutepow, N., Bittler, K., Neubauer, D., Reis, H. (BASF): Belg. Pat. 673943 (1966).
518. Chatt. J., Mann, F. G.: J. Chem. Soc. **1939**, 1631.
519. Bhattacharyya, S. K., Nag, S. N.: Brennstoff-Chem. **43**, 114 (1962).
520. Du Pont: Brit. Pat. 651853 (1951), C. A. **46**, 5615 (1952).
521. Kröper, H., v. Kutepow, N., Huchler, O., Kölsch, W., Himmele, W., (BASF): German Pat. 920244 (1954), Z. **1955**, 5896.
522. Tsuji, J., Morikawa, M., Kiji, J.: Tetrahedron Letters **1963**, 1437.
523. Natta, G., Pino, P., Ercoli, R., (Montecatini): U. S. Pat. 2805245 (1957), C. A. **52**, 6400 (1958).
524. Reppe, W., Kröper, H. (BASF): German Pat. 879987 (1953), Z. **1954**, 11047.
525. — — Liebigs Ann. Chem. **582**, 38—71 (1953).
526. Natta, G., Pino, P., Mantica, E.: Chim. Ind. (Milan) **32**, 201 (1950).
527. Brewis, S., Hughes, P. R.: Chem. Commun. **1965**, 157.
528. — — Chem. Commun. **1965**, 489.
529. Tsuji, J., Hosaka, S., Kiji, J., Suzuki, T.: Bull. Chem. Soc. Japan **39**, 141 (1966).
530. Reppe, W., Kröper, H.: Liebigs Ann., Chem. **582**, 38 (1953).
531. Alderson, T., Engelhardt, V. A. (Du Pont): U. S. Pat. 3065242 (1962), C. A. **58**, 8912 (1963).

532. v. Kutepow, N., Bittler, K., Neubauer, D., Reis, H. (BASF): German Pat. Appl. Pend.
533. Reppe, W., Kröper, H.: Liebigs Ann. Chem. 582, 61 (1953).
534. — — German Pat. 848355 (1952), Z. 1954, 8218.
535. — — German Pat. 862748 (1953), C. A. 48, 10059 (1954).
536. Brooks, R. E., Gresham, W. F., Hardy, J. V., Lupton, J. N.: Ind. Eng. Chem. 49, 2004 (1957).
537. Reppe, W., Kröper, H.: Liebigs Ann. Chem. 582, 69 (1953).
538. Pino, P., Magri, R.: Chim. Ind. (Milan) 34, 1 (1952).
539. Growe, B. F., Elmer, O. C. (Texas Co.): U. S. Pat. 2742502 (1956), C. A. 50, 16849 (1956).
540. Nienburg, H. J., Keunecke, E. (BASF): German Pat. 863799 (1941), C. A. 48, 1427 (1954).
541. Reppe, W., Kröper, H., Pistor, H. J. (IG Farben): German Pat. 763693 (1941), Z. 1954, 4486.
542. Hecht, O., Kröper, H.: Naturforschung und Medizin in Deutschland 1947, Vol. 36, Präparative organische Chemie, Part. I, p. 134. Ed. by K. Ziegler. Wiesbaden: Dieterich'sche Verlagsbuchhandlung 1948.
543. v. Kutepow, N., Himmele, W., Hohenschutz, H.: Chem. Ingr.-Tech. 37, 383 (1965).
544. Reppe, W., v. Kutepow, N., Bille, H., (BASF): U. S. Pat. 3014962 (1961), C. A. 57, 667 (1962).
545. Aliev, Y. Y., Isakov, Y. J.: Issled. Mineral'n. i Rast Syr'ya Uzbekistana, Akad. Nauk USSR, Inst. Khim. 1962, 95, C. A. 59, 3767 (1963).
546. Sourirajan, S., in: Advances in Catalysis, Vol. IX, p. 618. New York: Academic Press 1957.
547. Adkins, H., Rosenthal, R. W.: J. Am. Chem. Soc. 72, 4550 (1950).
548. Codignola, F., Piacenza, M.: Ital. Pat. 431407, C. A. 44, 1134 (1950).
549. BASF: Brit. Pat. 713515 (1954), C. A. 50, 6501 (1956).
550. Tsuji, J., Kiji, J., Morikawa, M.: Tetrahedron Letters 26, 1811 (1963).
551. Parshall, G. W.: Z. Naturforsch. 18b, 772 (1963).
552. Reppe, W., Kröper, H., Pistor, H. J., Weissbarth, O.: Liebigs Ann. Chem. 582, 90 (1953).
553. — — v. Kutepow, N. (BASF): German Pat. 879988 (1942), Z. 1955, 5651.
554. Eisenmann, J. L., Yamartino, R. L., Howard, jr., J. F.: J. Org. Chem. 26, 2102 (1961).
555. Reppe, W., Friedrich, H. (BASF): U. S. Pat. 2729651 (1956), C. A. 50, 13081 (1956).
556. Bhattacharyya, S. K., Vir, D., in: Advances in Catalysis, Vol. IX, p. 625. New York: Academic Press 1957.
557. Kato, J., Iwanaga, R., Wakamatsu, H. (Ajinomoto Co): German Pat. 1135884 (1962), C. A. 58, 4430 (1963).
558. Heck, R. F.: J. Am. Chem. Soc. 85, 2013 (1963).
559. Kröper, H.: Anlagerung von Kohlenmonoxyd und Wasserstoff (Hydroformylierung). In: Houben-Weyl, Vol. IV/2, p. 401, Stuttgart: Georg Thieme 1955.
560. Reppe, W., Kröper, H., Pistor, H. J., Weissbarth, O.: Liebigs Ann. Chem. 582, 105 (1953).
561. Hecht, O., Kröper, H.: Naturforschung und Medizin in Deutschland, Vol. 36, Präparative organische Chemie, Part. 1, p. 139. Ed. by K. Ziegler. Wiesbaden: Dieterich'sche Verlagsbuchhandlung 1948.

562. Chiusoli, G. P., Merzoni, S.: Z. Naturforsch. **17b**, 850 (1962).
563. Dent, W. T., Long, R., Whitfield, G. W.: J. Chem. Soc. **1964**, 1588.
564. Chiusoli, G. P.: Chim. Ind. (Milan) **41**, 513 (1959).
565. Rosenthal, R. W.: J. Org. Chem. **24**, 836 (1959).
566. Tsuji, J., Nogi, T.: Tetrahedron Letters **1966**, 1801.
567. Bauld, N. L.: Tetrahedron Letters No. **27**, 1841 (1963).
568. Yamamoto, K., Kato, S.: Jap. Pat. 2424, C. A. **48**, 2105 (1954).
569. Prichard, W. W., Tabet, G. E.: U. S. Pat. 2565462 (1951), C. A. **46**, 2578 (1952).
570. Kröper, H., Wirth, F., Huchler, O.: Angew. Chem. **22**, 867 (1960).
571. — — — German Pat. 1033654 (1955), C. A. **54**, 12069 (1960).
572. — — — German Pat. 1052974 (1955), Z. **1960**, 651.
573. — — — German Pat. 1062691 (1957), Z. **1960**, 7039.
574. — — — German Pat. 1066574 (1957), Z. **1960**, 8349.
575. — — — German Pat. 1074 028 (1957), Z. **1960**, 11820.
576. Bliss, H., Southworth, R. W.: U. S. Pat. 2565461, C. A. **46**, 2577 (1952).
577. Tabet, G. E.: U. S. Pat. 2565463, C. A. **46**, 2578 (1952).
578. Rohm & Haas: U. S. Pat. 2582911 (1952), C. A. **46**, 11231 (1952).
579. — U. S. Pat. 2582299 (1952), C. A. **46**, 8671 (1952).
580. Dakli, I., Arsizoi, B., Corsi, L. (Montecatini): U. S. Pat. 2881205 (1959), C. A. **53**, 15982 (1959).
581. Reppe, W., Reicheneder, F., Stengel, G., Ziegler, A. (BASF): U. S. Pat. 2925436 (1960), C. A. **54**, 17270 (1960).
582. — Chemie und Technik der Acetylen-Druck-Reaktionen, Weinheim: Verlag Chemie 1952.
583. Eur. Chem. News **131**, 26 (1964).
584. Sittig, M.: Acrylic acid and esters. In: Chem. Proc. Monograph. No. 13, p. 5—22, Park Bridge. New Jersey, Noyes Development Corp. 1965.
585. v. Kutepow, N., Himmele, W. (BASF): German Pat. 1026297, Z. **1958**, 14, 172.
586. Ford, T. A.: U. S. Pat. 2491131 (1947).
587. Roland, I. R., Wilson II, I. D. C., Hanford, W. E.: J. Am. Chem. Soc. **72**, 2122 (1956).
588. Koch, H., Gilfert, W.: Part of the work of H. Koch, in: Brennstoff-Chem. **36**, 321 (1955).
589. Huisken, W.: Diplomarbeit, Universität Bonn (1952).
590. Koch, H., Haaf, W.: Angew. Chem. **70**, 311 (1958).
591. — — Liebigs Ann. Chem. **618**, 251 (1958).
592. Haaf, W.: Brennstoff-Chem. **45**, 209 (1964).
593. Koch, H.: Fette, Seifen, Anstrichmittel **59**, 493 (1957).
594. Eidus, Ya. T., Puzitskii, K. W., Ryabova, K. G.: Proc. Acad. Sci. USSR **120**, 323 (1958).
595. Balaban, A. T., Nenitzescu, C. D.: Tetrahedron **10**, 55 (1960).
596. Koch, H., Möller, K. E. (Studienges. Kohle): German Pat. 1064941 (1960), Z. **1960**, 3380.
597. Anderson, J. E., Franke, N. W. (Gulf Research and Dev.): U. S. Pat. 3167585 (1965).
598. Friedman, B. S., Cotton, S. M.: J. Org. Chem. **27**, 481 (1962).
599. Möller, K. E.: Angew. Chem. **73**, 767 (1961).
600. — Brennstoff-Chem. **47**, 10 (1966).
601. Eidus, Ya. T., Kaal, T. A.: J. Gen. Chem. USSR **34**, 3447 (1964).
602. Koch, H.: Riv. Combust. **10**, 77 (1956).

603. Rohloffs, G., Pawlenko, S. (Schering AG): German Pat. 1177133 (1959).
604. Pawlenko, S. (Schering AG): German Pat. 1211621 (1961), C. A. **59**, 11264 (1966).
605. — German Pat. 1212061 (1961), C. A. **59**, 11264 (1963).
606. Friedman, B. S. (Sinclair Refining): U. S. Pat. 2874186 (1959), C. A. **54**, 14153 (1960).
607. Koch, H., Huisken, W. (Studienges. Kohle): German Pat. 972291, see Brit. Pat. 798065 (1958), C. A. **53**, 6083 (1959).
608. — — German Pat. 973077, see Brit. Pat. 798065 (1958), C. A. **53**, 6083 (1959).
609. Rohloffs, G., Pawlenko, St. (Schering AG): U. S. Pat. 3099687 (1963), C. A. **58**, 4429 (1963).
610. Shell: Fr. Pat. 1252675 (1965).
611. Möller, K. E.: Brennstoff-Chem. **45**, 129 (1964).
612. Hine, J.: Reaktivität und Mechanismus in der organischen Chemie, p. 303. Stuttgart: Georg Thieme 1960.
613. Möller, K. E.: Angew. Chem. **75**, 1098 (1963).
614. Koch, H., Huisken, W., Möller, K. E., Lohbeck, K.: Addition to German Pat. 942987 under St. 8534 IV/b/120 (1954).
615. — Brennstoff-Chem. **36**, 321 (1955).
616. — Hiusken, W. (Studienges. Kohle): German Pat. 942987 (1956), C. A. **52**, 16204 (1958).
617. Möller, K. E.: Diss., T. H. Aachen 1954.
618. Haaf, W.: Dipl.-Arb. Univ. Bonn 1955.
619. — Chem. Ber. **99**, 1149 (1966).
620. Schauerte, K. H.: Diss., T. H. Aachen 1962.
621. Schering AG: Brit. Pat. 908497 (1962), U. S. Pat. 3099687, C. A. **58**, 4429 (1963).
622. Shell: Brit. Pat. 883142 (1959), C. A. **56**, 8570 (1962).
623. Genas, M., Rull, T.: Bull. Soc. Chim. France **1962**, 1837.
624. Möller, K. E.: Angew. Chem. **75**, 1122 (1963).
625. Vos, I. M., de Vries, R. (Shell): U. S. Pat. 3059007 (1962), C. A. **58**, 5519 (1963).
626. Koch, H., Möller, K. E.: Angew. Chem. **73**, 240 (1961).
627. — (Studienges. Kohle): German Pat. 942987 (1956), Z. **1957**, 797.
628. — Möller, K. E.: German Pat. 1095802 (1958), C. A. **56**, 8570 (1962).
629. Anderson, J. E., Franke, N. F. (Gulf Research): U. S. Pat. 3167585 (1965), C. A. **62**, 7642 (1965).
630. Koch, H.: U..S. Pat. 3061621 (1962), C. A. **58**, 4430 (1963).
631. Möller, K. E.: Brennstoff-Chem. **45**, 129 (1964).
632. Koch, H., Haaf, W.: Liebigs Ann. Chem. **638**, 111 (1960).
633. Studien u. Verwertungsges. Mülheim: Fr. Pat. 1076357 (1954), Z. **1956**, 7363.
634. Roe, E. T., Swern, D.: J. Am. Oil Chemist's Soc. **37**, 661 (1960).
635. Weintraub, L., Vitcha, J. F., Limon R.: Chem. Ind. (London) **1965**, 185.
636. Falbe, J., Paatz, R., Korte, F.: Chem. Ber. **97**, 3088 (1964).
637. Paatz, R., Korte, F., Weisgerber, G.: Chem. Ber. **100**, 984 (1967).
638. Schneider, A. (Sun Oil): U. S. Pat. 2864858 (1958), C. A. **53**, 9063 (1959).
639. — U. S. Pat. 2864859 (1958), C. A. **53**, 9156 (1959).
640. Haaf, W., Koch, H.: Liebigs Ann. Chem. **638**, 122 (1960).
641. Puzitskii, K. V., Eidus, Ya. T., Ryabova, K. G.: Zh. Obshch. Khim. **33**, 3278 (1963).

642. Koch, H., Haaf, W.: Angew. Chem. **72**, 628 (1960).
643. — — Chem. Ber. **94**, 1252 (1961).
644. Christol, H., Laurent, A., Mousseron, M.: Bull. Soc. Chim. France **23**, 19 (1961).
645. Lasser, B. T.: Diss., T. H. Aachen 1962.
646. Takezaki, Y., Kawatani, T., Sugita, N., Osugi, M., Yuasa, S., Suzuki, Y.: Bull. Japan Petrol. Inst. **2**, 94 (1960).
647. Eidus, Ya. T., Puzitskii, K. V., Guseva, I. V.: Zh. Obshch. Khim. **32**, 2983 (1963).
648. Puzitskii, K. V., Eidus, Ya. T., Ryabova, K. G.: Dokl. Akad. Nauk SSSR **141**, 636 (1961).
649. Eidus, Ya. T., Kaal, T. A.: Zh. Obshch. Khim. **35**, 120 (1965).
650. Takezaki, Y.: Kogyo Kagaku Zasshi **60**, 1038 (1957), C. A. **53**, 13046 (1959).
651. Haaf, W.: Chem. Ber. **99**, 1149 (1966).
652. Benedictis, A. D., Furman, K. E. (Shell): U. S. Pat. 2913489, C. A. **55**, 19790 (1961).
653. Himmele, W. (BASF): Unpublished.
654. Christol, H., Solladie, G.: Bull. Soc. Chim. France **1966**, 1307.
655. Eidus, Ya. T., Kaal, T. A.: J. Gen. Chem. USSR **35**, 119 (1965).
656. Haaf, W.: Chem. Ber. **96**, 3359 (1963).
657. Chem. Ind. Düsseldorf **62**, 381 (1948).
658. Loder, D. J. (Du Pont): U. S. Pat. 2152852 (1939), C. A. **33**, 5006 (1939).
659. Larsen, A. T. (Du Pont): U. S. Pat. 2153064 (1939), C. A. **33**, 5006 (1939).
660. Himmele, W. (BASF): German Pat. 1192178 (1965).
661. — German Pat. 1227010 (1966).
662. — Private Communication.
663. Möller, K. E.: Brennstoff-Chem. **47**. 15 (1966).
664. v. Dam, J., Waale, M. J.: Chim. Ind. **90**, 511 (1963).
665. Herzberg, S.: VI. FATIPEC-Kongress (1962), p. 319, Weinheim: Verlag Chemie 1962.
666. Goppel, J. M., Bruin, P., Zonsfeld, J. J.: Farbe Lack **69**, 181 (1963).
667. Bruin, P., Oosterhaf, H. A., Vegter, G. C., Vogelzang, E. J. W.: VII. FATIPEC-Kongress (1964), p. 49—60. Weinheim: Verlag Chemie 1964.
668. Falbe, J.: Angew. Chem. **78**, 532, (1966); Angew. Chem. Intern. Ed. Engl. **5**, 435 (1966).
669. — Schulze-Steinen, H. J.: Unpublished.
670. Holmquist, H. E., Carnahan, J. E.: J. Org. Chem. **25**, 2240 (1960).
671. Falbe, J.: Unpublished.
672. — Korte, F.: Brennstoff-Chem. **46**, 276 (1965).
673. — — Chem. Ber. **95**, 2680 (1962).
674. — — Chem. Ber. **98**, 1928 (1965); Falbe, J., Weitkamp, H., Korte, F.: Tetrahedron Letters **31**, 2677 (1965).
675. — — Angew. Chem. **74**, 900 (1962); Angew. Chem. Intern. Ed. Engl. **1**, 657 (1962).
676. Horiie, S., Murahashi, S.: Bull. Chem. Soc. Japan **33**, 247 (1960).
677. Rosenthal, A., Gervay, J.: Can. J. Chem. **42**, 1490 (1964).
678. Horiie, S., Murahashi, S.: Bull. Chem. Soc. Japan **33**, 88 (1960).
679. Brewis, S., Hughes, P. R.: Chem. Commun. **1966**, 6.
680. Chiusoli, G. P., Botaccio, E. G.: Chim. Ind. (Milan) **44**, 1129 (1962).
681. Falbe, J., Korte, F.: Abstracts, IUPAC Kongress. London: 1962.
682. — — Chem. Ingr.-Tech. **36**, 158 (1964).

683. — — Angew. Chem. **74**, 291 (1962), Angew. Chem. Intern. Ed. (Engl.) **1**, 266 (1962).
684. — Schulze-Steinen, H.-J., Korte, F.: Chem. Ber. **98**, 1923 (1965).
685. Nicholson, J. K., Shaw, B. L.: Proc. Chem. Soc. **1963**, 282.
686. Prichard, W. W. (Du Pont): U. S. Pat. 2841591 (1958), C. A. **52**, 20197 (1958).
687. Murahashi, S., Horiie, S.: Ann. Rep. Sci. Works, Fac. Sci. Osaka Univ. **7**, 89 (1959).
688. Rosenthal, A., Gervay, J.: Chem. Ind. (London) **1963**, 1623.
689. Zilbermann, E. N., Kalugin, A. A., Perepletchikova, E. M.: J. Gen. Chem. USSR **1962**, 900, C. A. **58**, 1337 (1963).
690. Rosenthal, A., Astbury, R. F., Hubscher, A.: J. Org. Chem. **23**, 1037 (1958).
691. — Weir, M. S.: Can. J. Chem. **40**, 610 (1962).
692. — Yalpani, M.: Can. J. Chem. **43**, 3449 (1965).
693. — Millward, S.: Can. J. Chem. **41**, 2504 (1963).
694. O'Connor, R.: J. Org. Chem. **26**, 4375 (1961).
695. Borsche, W., Merkwitz, C.: Ber. **37**, 3180 (1904).
696. Murahashi, S., Horii, S.: J. Am. Chem. Soc. **78**, 4816 (1956).
697. Horii, S., Murahashi, S.: Bull. Chem. Soc. Japan **33**, 247 (1960).
698. Kim, P., Hagihara, N.: Bull. Chem. Soc. Japan **38**, II, 2022.
699. Klemchuk, P. P.: U. S. Pat. 2995607 (1961).
700. Heck, R. F.: J. Am. Chem. Soc. **85**, 3116 (1963).
701. Mullineuax, R. D. (Shell): Belg. Pat. 603820 (15. 1. 1961).
702. Slaugh, L. H., Mullineaux, R. D. (Shell): Belg. Pat. 606408 (1961), U. S. Pat. 3239569 (1966).
703. — Belg. Pat. 619344 (1962).
704. Greene, C. R., Meeker, R. E. (Shell): Belg. Pat. 621833 (1962), C. A. **59**, 11259 (1963); U. S. Pat. 3274263 (1966).
705. — Belg. Pat. 623213 (1962), C. A. **60**, 15732 (1964).
706. — Belg. Pat. 627365 (1963), C. A. **60**, 9149 (1964).
707. — Belg. Pat. 627371 (1963), C. A. **60**, 6746 (1964).
708. Mertzweiller, J. K. (Esso): U. S. Pat. 3094564 (1960).
709. — Watts, R. N. (Esso): German Pat. 1114469 (1960), Z. **1962**, 10636.
710. Aldridge, C. L. (Esso): U. S. Pat. 2963449 (1958).
711. Cull, N. L. (Esso): German Pat. 1076654 (1958), Z. **1961**, 299.
712. Esso: German Pat. 1024943 (1955), C. A. **51**, 10859 (1960).
713. — Brit. Pat. 761024, C. A. **51**, 15548 (1957).
714. Aldridge, C. L. (Esso): U. S. Pat. 2942034 (1957), C. A. **54**, 34394 (1960).
715. Cull, N. L., Aldridge, C. L., Mertzweiller, J. K. (Esso): U. S. Pat. 2845465 (1958), C. A. **53**, 228 (1959).
716. — — — (Esso): Brit. Pat. 867799 (1959).
717. Esso: U. S. Pat. 2811567, C. A. **52**, 4677 (1958).
718. Chem. Eng. **68**, 25, 70—71 (1961).
719. Vickers, E. J., Reynolds, R. J. W. (I. C. I.): Brit. Pat. 775495 (1957), C. A. **52**, 1210 (1958).
720. Esso: German Pat. 1011410 (1954), C. A. **53**, 15974 (1959).
721. Belg. Pat. 574599 (1959).
722. I. C. I. (Fober): **46**, No. 9, 1206.
723. Taylor, A. W. C. (I. C. I.): German Pat. 1027194 (1958), C. A. **54**, 11992 (1960).
724. I. C. I.: Fr. Pat. 1224115 (1959), Z. **1962**, 13942.
725. Carpenter, G. B. (Du Pont): U. S. Pat. 1924763 (1932), Z. **1933**, 3194.

726. — U. S. Pat. 1924766 (1932), Z. **1933**, 3193.
727. — U. S. Pat. 1924767 (1932), Z. **1933**, 3193.
728. — U. S. Pat. 1957939 (1934), Z. **1934**, 1370.
729. Larson, A. T., Vail, W. E. (Du Pont): U. S. Pat. 1924765 (1932), Z. **1933**, 3194.
730. Vail, W. E. (Du Pont): U. S. Pat. 1924764 (1932), Z. **1933**, 3194.
731. Woodhouse, J. C. (Du Pont): U. S. Pat. 1924762 (1932), C. A. **27**, 5339 (1933).
732. Notwendige Schritte Deutscher Technik, Heft 3 (1953); Sonderheft: Industrielle Verwertung des Kohlenoxyds, by Prof. Dr. W. Fuchs. Droste Verlag, Düsseldorf.
733. Osumi, Y., Yamaguchi, H., Onoda, T., Onishi, M. (Mitsubishi Chem. Ind. Co. Ltd.): Jap. Pat. 22735/65 (1965), C. A. **64**, 4943 (1966).
734. Ajinomoto Co. Inc.: Jap. Pat. 1419/64 (1964).
735. Morikawa, M.: Bull. Chem. Soc. Japan **37**, 379—380 (1964).
736. Ajinomoto Co. Inc.: Jap. Pat. 1575/65 (1965).
737. Cannel, L. G., Slaugh, L. H. (Shell): German Pat. 1186455 (1965), C. A. **62**, 16054 (1965).
738. Ajinomoto Co. Inc.: Jap. Appl. 575/66 (1966).
739. Mitsubishi Chem. Ind.: Jap. Appl. 653/66 (1966).
740. Jap. Appl. 207/66 (1966): Fortschrittsber. **52**, 22191 (1966).
741. Rehner, jr., J. (Esso): German Pat. 1093347, U. S. Pat. 3003938.
742. Slaugh, L. H., Mullineaux, R. D. (Shell): U. S. Pat. 3239571 (1966), C. A. **65**, 618 (1966).
743. — — U. S. Pat. 3239566 (1966), C. A. **64**, 15745 (1966).
744. Klempt, W. (Bergwerksverband zur Verwertung von Schutzrechten der Kohlentechnik GmbH): German Pat. 710170 (1941), Z. **1941**, 2869, U. S. Pat. 2153852, Fr. Pat. 831474.
745. Shattuck, M. T. (Du Pont): U. S. Pat. 2443482 (1948), C. A. **42**, 7324 (1948).
746. Johnson, R. (Du Pont): U. S. Pat. 2273269 (1942), C. A. **36**, 3810 (1942).
747. Gresham, W. F. (Du Pont): U. S. Pat. 2364438 (1944), C. A. **39**, 4632 (1945).
748. Europa Chemie **24**, 9 (1966).
749. Standard Oil: German Pat. 944728 (1965), C. A. **53**, 3056 (1959).
750. Dubini, M., Chiusoli, G. P., Montino, F.: Tetrahedron Letters **1963**, 1591.
751. Koontz, J. D. (Standard Oil): German Pat. 1057593 (1952), U. S. Pat. 2679534 (1954), C. A. **48**, 10329 (1954).
752. Mertzweiler, J. K. (Esso): U. S. Pat. 2841617 (1958), C. A. **52**, 17109 (1958).
753. Munger, S. H. (Du Pont): U. S. Pat. 2779796 (1957), C. A. **51**, 9674 (1957).
754. Mertzweiler, J. K., Templeton, H. E., Daussat, R. L. (Esso): Brit. Pat. 735352 (1952), Z. **1957**, 4819.
755. Gwynn, B. H. (Gulf Research): Brit. Pat. 785991 (1957), Z. **1959**, 15504.
756. Natta, G., Beati, E.: Brit. Pat. 646424 (1950), C. A. **45**, 5714 (1951).
757. Esso: Brit. Pat. 912974 (1962), Z. **1964**, 44—2101.
758. Aldridge, C. L., Cull, N. L. (Esso): Brit. Pat. 907027 (1962), Z. **1964**, 47—2250.
759. Heck, R. F.: Advanc. Organometal. Chem. **4**, 243 (1966).
760. Almasy, M., Szabo, L.: Acad. Rep. Populare Romine, Studii Cercetari Chim. **8**, 531 (1960).

198   References

761. Marko, L.: Ber. Ungarische Mineralöl- und Erdgas-Versuchsanstalt, **2**, 228 (1961).
762. Noyori, G., Honda, M., Koga, T. (Zaindan-Hogin Noguchi Keukyusho): Jap. Pat. 41—6734 (1956).
763. Roelen, O.: FIAT Rev. Ger. Sci. 1939—1946, Part I, p. 167. Wiesbaden: Dieterich'sche Verlagsbuchhandlung 1948.
764. Max, N. (Shell): Fr. Pat. 984772 (1951), Swedish Pat. 130809 (1951), Z. **1952**, 2256.
765. Gankin, W. J.: USSR Pat. 137509 (1960), Z. **1964**, 2410.
766. Esso: Brit. Pat. 690977 (1951), Z. **1955**, 7781.
767. — U. S. Pat. 2757200 (1956), Z. **1957**, 7228.
768. BASF: Pat. Appl. B 21745 (1952).
769. Esso: German Pat. 1138754 (1962), Z. **1964**, 2388.
770. Gulf Research: Fr. Pat. 1371603 (1964), C. A. **62**, 3692 (1965).
771. California Research Corp.: U. S. Pat. 2547178 (1948), Z. **1954**, 1564.
772. Standard Oil: German Pat. 977576 (1956).
773. Field, E. (Standard Oil): U. S. Pat. 2683177 (1954), Z. **1955**, 6848.
774. Gulf Publishing Company Publication, Petrol Refiner **31**, 3, 149—150 (1952).
775. Standard Oil: Brit. Pat. 928905 (1963), Z. **1965**, 2—2651.
776. Agency of Industrial Science and Technology: Jap. Pat. 38—2986 (1963).
777. Guccione, E.: Chem. Eng. **72**, 90—92 (1965).
778. Asinger, F.: Chemie und Technologie der Monoolefine, p. 700. Berlin; Akademie-Verlag 1957.
779. Reppe, W., Schlenk, W. (BASF): German Pat. 753618.
780. Meis, J., Tummes, H., (Ruhrchemie AG): German Pat..1235285 (1967).
781. Ramp, F. L., Dewitt, E. J., Trapaso, L. E.: J. Polymer Sci., Part. A-1, **4**, 2267 (1966).
782. Cull, N. L., Aldridge, C. L. (Esso): U. S. Pat. 3118948 (1964), C. A. **60**, 9149 (1964).
783. — — (Esso): U. S. Pat. 2908721 (1959), Z. **1961**, 12644.
784. Jaros, S. E., Roming, jr., C. (Esso): U. S. Pat. 3119876 (1964), Z. **1965**, 21—2705.
785. Cull, N. L., Aldridge, C. L. (Esso): U. S. Pat. 2862979 (1958), Z. **1961**, 2439.
786. Kato, J., Ito, T., Yabe, Y.: Kogyo Kagaku Zasshi **65**, 184 (1962), C. A. **58**, 4420 (1963).
787. Faith, W. L., Keyes, D. B., Clark, R. L.: Industrial Chemicals, 3rd Ed., p. 309, New York-London-Sydney: John Wiley & Sons 1965.
788. Chem. Ind. Düsseldorf **1964**, 636.
789. Guccione, E.: Chem. Eng. **1965**, 90—92.
790. Greene, C. R., Meeker, R. E. (Shell): German Pat. 1212953 (1962), C. A. **59**, 11259 (1963).
791. Cannell, L. G., Slaugh, L. H., Mullineaux, R. D. (Shell): German Pat. 1186455 (1965), Z. **1965**, 40—2557.
792. Ruhrchemie AG: Previously unpublished.
793. Lamola, A. A., (Du Pont): German Pat. 1219400 (1963).
794. Vickers, E. J., Reynolds, R. J. W. (I. C. I.): Brit. Pat. 775122 (1957), C. A. **51**, 17236 (1957).
795. Tsuda, T., Shimizu, T., Yamashita, Y.: Kogyo Kagaku Zasshi **67**, (10), 1661 (1964).
796. Ellis, W. J., Ronning, C.: Hydrocarbon Process. Petrol. Refiner **44**, 139 (1965).

797. Rosenthal, A., Wender, I.: Organic Syntheses via Metal Carbonyls, Vol. I, New York: p. 373. Interscience Publishers 1968.
798. Kim, P. J., Hagihara, N.: Mem. Inst. Sci. Ind. Res. Osaka Univ. 24, 133 (1967).
799. Chiusoli, G. P., Cassar, L.: Angew. Chem. Intern. Ed. 6, 124 (1967).
800. Cassar, L., Chiusoli, G. P.: Tetrahedron Letters 25, 1805 (1966).
801. — — Foa, M.: Chim. Ind. (Milan) 50, 515 (1968).
802. Himmele, W., (BASF): German Pat. 1240848 (1963).
803. Sahni, R. C.: Trans. Faraday Soc. 49, 1246 (1353).
804. Smeeton Leah A., in: Din, F., Thermodynamic Function of Gases, Vol. 1. London: Butterworth & Co. Ltd. 1956.
805. Greene, R. V., in: Kirk-Othmer, Encyclopedia of chemical technology, 2nd Ed., Vol. 4, p. 424. New York-London-Sydney: Interscience Publishers J. Wiley & Sons. Inc., 1964.
806. Esso: Brit. Pat. 1072796 (1964).
807. Ramp, F. L., Dewitt, E. J., Trapasso, L. E.: J. Polymer Sci., Part A-1, 4, 2267 (1966).
808. Falbe, J., in: Ullmanns Encyklopädie der technischen Chemie, Ergänzungsband, 3rd edit., p. 87. München-Berlin-Wien: Urban & Schwarzenberg 1970.
809. — Cornils, B.: Unpublished results.
810. Takegami, Y., Watanabe, Y., Masada, H.: Bull. Chem. Soc. Japan 40, 1459 (1967).
811. Heil, B., Marko, L.: Chem. Ber. 101, 2209 (1968).
812. Imyanitov, N. S., Rudkowskii, D. M.: Neftekhimiya 3, 198 (1963).
813. Gankin, W. J., Genender, L. S., Rudkowskii, D. M.: J. Appl. Chem. USSR 40, 2029 (1967).
814. Weigert, W. (Degussa): Neth. Pat. Appl. 6516193 (1966), C. A. 66, no. 2215 o, p. 217 (1967).
815. — (Degussa): German Pat. 1280237 (1965).
816. Bartlett, J. H., Hughes, V. L. (Esso): U. S. Pat. 2894038 (1959), C. A. 54, 2216a (1960).
817. Evans, D., Osborn, J. A., Wilkinson, G.: J. Chem. Soc. A 12/1968, 3133.
818. Brewis, S., Hughes, P. R.: Chem. Commun. 2, 71 (1967).
819. Nienburg, H. J., Helms, A., Pistor, H. J. (Chem. Verwertungsges. Oberhausen): German Pat. 914375 (1954).
820. Schuster, C., Eilbracht, H. (Chem. Verwertungsges. Oberhausen): German Pat. 892287 (1953).
821. Büchner, K., Kühnel, P. J. (Ruhrchemie AG): German Pat. 854216 (1952).
822. Schuster, C., Eilbracht, H. (Chem. Verwertungsges. Oberhausen): German Pat. 877300 (1953).
823. Gankin, W. J., Krinkin, D. P., Rudkowskii, D. M., in: Carbonylation of Unsaturated Hydrocarbons, p. 45 ff. Ed. by Allunions Sci. Res. Inst. for Petrochemical Processes. Leningrad: Chemistry Dept. 1968.
824. Imyanitov, N. S., Rudkowskii, D. M., in: Carbonylation of Unsaturated Hydrocarbons, p. 28 ff. Ed. by Allunions Sci. Res. Inst. for Petrochemical Processes. Leningrad: Chemistry Dept. 1968.
825. Falbe, J., Tummes, H., Weber, J.: Brennstoff-Chem. 50, 46 (1969).
826. — — — Brennstoff-Chem., 50, 20 (1969).
827. Tucci, E. R.: Ind. Eng. Chem. Prod. Res. Develop. 7, 125 (1968).
828. Fell, B., Rupilius, W., Asinger, F.: Tetrahedron Letters 29, 3261 (1968).
829. Barthory, J., Freund, M., Laky, L., Marko, L., Monostory, A.: Austr. Pat. 256799 (1967).
830. Heck, R. F.: Private communication to the author.

200 References

831. Basolo, F., Brault, A. T., Poe, A. J.: J. Chem. Soc. **1964**, 676.
832. Heck, R. F.: J. Am. Chem. Soc. **85**, 657 (1963).
833. — J. Am. Chem. Soc. **87**, 2572 (1965).
834. Werner, H.: Angew. Chem. **24**, 1017 (1968).
835. Matsuda, A.: Bull. Chem. Soc. Japan **41**, 1876 (1968).
836. Glasebrook, A. L.: Autoclave engineers technical bulletin no. **100**, Erie, Pa. (1957).
837. Tsiklis, D. S.: Handbook of techniques in high pressure research and engineering, p. 291. New York: Plenum Press 1968.
838. Büchner, K., Meis, J. (Ruhrchemie AG): German Pat. 1024499 (1958), C. A. **54**, 11999b (1960).
839. Carter, C. E. (Gulf Oil Corp.): Brit. Pat. 779388 (1953).
840. Nienburg, H. J., Eckert, E., Kölsch, W., Goilav, M., Pistor, H. J. (Chem. Verwertungsges. Oberhausen): German Pat. 953606 (1956), C. A. **53**, 5123 (1959).
841. Lemke, H. (Kuhlmann): Fr. Pat. 1089983 (1953), C. A. **53**, 12528 (1959).
842. Johnson, P. J., Cox, N. R., (Union Carbide Corp.): U. S. Pat. 3014970 (1957), C. A. **56**, 8567 (1962).
843. Lemke, H.: Hydrocarbon Process. Petrol Refiner **45**, 148 (1966).
844. Moell, H., Eckert, E., Kerber, H., Appl, M., Hohenschutz, H., Walz, H. (BASF): German Pat. 1272911 (1966).
845. Mistrik, E. J., Rendo, T.: Chem. Techn. (Berlin) **19**, 154 (1967).
846. Habeshaw, J., Rae, R. W. (Anglo-Iranian Oil): German Pat. 921934 (1952), Z. **1955**, 11094.
847. Tummes, H., Meis, J. (Ruhrchemie AG): German Pat. 1258855 (1965).
848. Rehn, K., Theiling, G. (Farbwerke Hoechst): German Pat. 1061770 (1957), C. A. **55**, 11305 (1961).
849. Vander Woude, J. C., Morris, P. M. (Eastman Kodak): U. S. Pat. 2763693 (1951), C. A. **51**, 8777 (1957).
850. Esso: Fr. Pat. 1420640 (1964), C. A. **65**, 10496 (1966).
851. Bartlett, J. H., Kirshenbaum, I., Muessig, C. W. (Esso): Ind. Eng. Chem. **51**, 257 (1959).
852. Falbe, J., in: Ullmanns Encyklopädie der technischen Chemie, Ergänzungsband 3rd edit., p. 98. München-Berlin-Wien: Urban & Schwarzenberg 1970.
853. Publication of Hoechst-Ruhrchemie: Produkte aus der Oxosynthese, 1969.
854. Falbe, J., Cornils, B.: Lösungen und Lösungsmittel. In: Fortschritte der chemischen Forschung, Bd. 11, No. 1, p. 101. Berlin-Heidelberg-New York: Springer 1968.
855. Hatch, L. F.: Higher oxo alcohols. New York: J. Wiley & Sons. 1957.
856. Lindner, K.: Tenside — Textilhilfsmittel — Waschrohstoffe. Stuttgart: Wissenschaftl. Verlagsges. m. b. H. 1964.
857. Heck, R. F., Breslow, D. S.: Actes 2e Congr. Intern. Catalyse, Paris 1960, p. 171. Paris: Technip. 1961.
858. Takegami, Y., Watanabe, Y., Masada, H., Mitsudo, T.: Bull. Chem. Soc. Japan, **42**, 206 (1969).
859. Eur. Chem. News **12**, no. 277, 21 (1967).
860. J. Commer. July 2nd, (1968).
861. Chem. Eng. News **46**, no. 30, 17 (1968).
862. Chem. Ind. Düsseldorf **XX**, 733 (1968).
863. Chem. Ind. Düsseldorf, **XIX**, 744 (1967).
864. Chem. Ind. Düsseldorf, **XX**, 8, 519, (1968).
865. Chemical Marketing Newspaper Dec. 2, (1968).

866. Eur. Chem. News **15**, 8 (1969).
867. Oil Gas Intern. **7**, 98 (1969).
868. Chem. and Ind. no. **6**, 161 (1968).
869. Eur. Chem. News no. **283**, 22 (1967).
870. Chem. Week, **100**, 30 (1967).
871. Eur. Chem. News no. **275**, 24 (1967).
872. Chem. Ind. Düsseldorf **XXI**, 257 (1969).
873. Europa-Chemie **4**, 12 (1969).
874. Oel-Zeitschrift für die Mineralölwirtschaft **7**, no. 4, 135 (1969).
875. Chem. Eng. Vol. **76**, 71 (1969).
876. Hydrocarbon Process. Petrol. Refiner Vol. **48**, no. 5, 58-G (1969).
877. Chem. Week May **20**, 63 (1967).
878. VWD-Chemie **141**, 1, (1969).
879. VWD-Chemie, **96**, 3, (1969).
880. Eur. Chem. News **374**, 38, (1969).
881. Europa-Chemie **7**, 4 (1968).
882. Europa-Chemie **6**, 12 (1968).
883. The Chemical Daily (Japan), Jan. 23, (1969).
884. The Weekly Chemical (Japan) Jan. 20, (1969).
885. The Petrochemical (Japan) Nov. 6, (1967).
886. Eur. Chem. News **377**, 8 (1969).
887. Repening, K.: Oel-Zeitschrift für die Mineralölwirtschaft, **11**, 352 (1968).
888. Chem. Eng. News no. 17, 26, (1968).
889. Chem. Eng. News no. 19, 28 (1969).
890. Chem. Ind. Düsseldorf **XX**, 537 (1968).
891. Kyle, H. E.: Oxo-Process, in: Kirk-Othmer, Encyclopedia of Chemical Technology, Vol. 14, 2nd Edition, p. 383. New York-London-Sydney: Interscience Publishers, J. Wiley & Sons, Inc. 1967.
892. Slaugh, L. H., Mullineaux, R. D.: J. Organometal. Chem. **13**, 469 (1968).
893. Guccione, E.: Chem. Eng. **72**, 90 (1965).
894. Iwasaki, I., Yumura, S.: Jap. Chem. Quart. **2**, (1), 71 (1966).
895. VWD-Chemie, **226**, 6 (1968).
896. VDI-Nachrichten **1**, 4 (1969).
897. G. I. T.-Fachz. Lab. **13**, 560, (1969).
898. Chem. Ind. Düsseldorf **XX**, 822 (2968).
899. Hill, M.: Hydrocarbon Process. Petrol. Refiner **43** (8), 135 (1964).
900. Eur. Chem. News no. **290**, 18 (1967).
901. Gankin, W. J., Gordina, N. J., Krinkin, D. P., Rudkowskii, D. M., Trifel, A. G.: Zh. Prikl. Khim. **40**, 1639 (1967).
902. Chem. Ind. Düsseldorf **XX**, 750 (1968).
903. Jap. Chem. Week June 20, p. 7 (1968).
904. Ferguson, G. U., in: The Production of Polymer and Plastic Intermediates from Petroleum, p. 86. By R. Long. London: Butterworth & Co. Ltd. 1967.
905. Roelen, O.: Über die Aldehyd-Reaktionen von Olefinen mit Kohlenmonoxid und Wasserstoff und verwandten Reaktionen, paper presented during the session of the German Chemical Soc. on Sept. 28, 1951 in Cologne.
906. Wisokinskii, G. P., Gankin, W. J., Rudkowskii, D. M., in: Carbonylation of Unsaturated Hydrocarbons, p. 17. Edited by Allunions Sci. Res. Inst. for Petrochemical Processes. Leningrad: Chemistry Dept. 1968.
907. Nagy-Magos, Z., Bor, G., Marko, L.: J. Organometal. Chem. **14**, 205 (1968).
908. Noack, K., Calderazzo, F.: J. Organometal. Chem. **10**, 101 (1967).

909. Coffield, T. H., Kozikowski, J., Closson, R. D.: J. Chem. Soc. Spec. Publ. no. 13, p. 126 (1959).
910. Mawby, R. J., Basolo, F., Pearson, R. G.: J. Am. Chem. Soc. **86**, 3994 (1964).
911. Falbe, J.: Lecture to session of the German Chemical Soc., Berlin 12-2-68, Abstr. in Angew. Chem. **80**, 568 (1968).
912. Kniese, W., Nienburg, H. J., Fischer, R.: J. Organometal. Chem. **17**, 133 (1969).
913. Marko, L.: Acta Chim. Acad. Sci. Hung. **59**, 396 (1969).
914. Johnson, M.: Chem. Ind. (London) **1963**, 684.
915. Manual, T. A.: J. Org. Chem. **27**, 3941 (1962).
916. Chalk, A. J., Harrod, J. F.: Advanc. Organometal. Chem. **6**, 119—170 (1968).
917. Takegami, Y., Yokokawa, C., Watanabe, Y., Masada, H., Okuda, Y.: Bull. Chem. Soc. Japan **37**, 1190 (1964).
918. — — — Okuda, Y.: Bull. Chem. Soc. Japan **37**, 181 (1964).
919. Piacenti, F., Pucci, S., Bianchi, M., Lazzaroni, R., Pino, P.: J. Am. Chem. Soc. **90**, 6847 (1968).
920. Rosenthal, A., Abson, D.: Can. J. Chem. **42**, 1811 (1964).
921. — Koch, H. J.: Can. J. Chem. **43**, 1375 (1965).
922. Schweckendiek, L.: German Pat. 841589 (1952).
923. v. Kutepow, N., Bille, H., (BASF): German Pat. 921988 (1955).
924. Lautenschlager, H., Friedrich, H.: German Pat. 1046030 (1959).
925. Diamond Alkali Comp.: Belg. Pat. 587621 (1960).
926. Schweckendiek, W.: German Pat. 834991 (1952).
927. Falbe, J.: Oxosynthese. In: Ullmanns Encyklopädie der technischen Chemie, Ergänzungsband, p. 87 (1969). München-Berlin: Urban & Schwarzenberg 1969.
928. Ruhrchemie AG: German Pat. Appl. unpublished.
929. Evans, D., Osborn, J. A., Wilkinson, G.: J. Chem. Soc. A **1968**, 3133.
930. Brown, C. K., Wilkinson, G.: Tetrahedron Letters **22**, 1725 (1969).
931. Pruett, R. L., Smith, J. A.: J. Org. Chem. **34**, 327 (1969).
932. Slaugh, L. H., Mullineaux, R. D.: J. Organometal. Chem. **13**, 469 (1968).
933. Evans, D., Yagupsky, G., Wilkinson, G.: J. Chem. Soc. A **1968**, 2660.
934. Tucci, E. R.: Ind. Eng. Chem. Prod. Res. Develop. **7**, 32 (1968).
935. — Ind. Eng. Chem. Prod. Res. Develop. **7**, 125 (1968).
936. Hershman, A., Graddock, J.: Ind. Eng. Chem. Prod. Res. Develop. **7**, 226 (1968).
937. Tucci, E. R.: Ind. Eng. Chem. Prod. Res. Develop. **7**, 227 (1968).
938. Osborn, J. A.: Endeavour **26**, 144 (1967).
939. Piacenti, F., Bianchi, M., Benedetti, E.: Chim. Ind. (Milan) **49**, 245 (1967).
940. Ibers, J. A.: J. Organometal. Chem. **14**, 423 (1968).
941. Simon, A., Nagy-Magos, Z., Palagyi, J., Palyi, G., Bor, G., Marko, L.: J. Organometal. Chem. **11**, 634 (1968).
942. Angelici, R. J.: Organometal. Chem. Rev. **3**, 173—226 (1968).
943. Fell, B., Rupilius, W., Asinger, F.: Tetrahedron Letters **29**, 3261—3266 (1968).
944. Falbe, J.: Synthesen mit Kohlenmonoxyd, Organische Chemie i. E., p. 19. Berlin-Heidelberg-New York: Springer 1967.
945. Sacco, A., Freni, M.: Ann. Chim. (Rome) **48**, 218 (1958).
946. Hieber, W., Freyer, W.: Chem. Ber. **93**, 462 (1960).
947. Orgel, L. E.: Introduction to Transition-Metal Chemistry, p. 132. New York: Wiley 1960.

948. Hieber, W., Linder, E.: Chem. Ber. **94**, 1417 (1961).
949. Bath, S. S., Vaska, L.: J. Am. Chem. Soc. **85**, 3500 (1963).
950. LaPlaca, S. J., Ibers, J. A.: J. Am. Chem. Soc. **85**, 3501 (1963).
951. Yamaguchi, M.: Kogyo Kagaku Zasshi **72**, 671 (1969).
952. Guccione, E.: Chem. Eng. Vol. **72**, 90 (1965).
953. Marko, L., Freund, M.: Acta Chim. Acad. Sci. Hung. **57** (4), 445 (1968).
954. — Acta Chim. Acad. Sci. Hung. **59**, 389 (1969).
955. Klumpp, E., Bor, G., Marko, L.: J. Organometal. Chem. **11**, 207 (1968).
956. Heil, B., Marko, L.: Chem. Ber. **102**, 2238 (1969).
957. Brewis, S. (I. C. I.): German Pat. 1273518 (1968).
958. Falbe, J., Tummes, H., Meis, J. (Ruhrchemie AG): German Pat. Appl. 1295537 (1969).
959. — Weber, J. (Ruhrchemie AG): German Pat. Appl. 1290535 (1968).
960. Gunter, C. G., Baldner, R. L., Wennerberg, A. N.: German Pat. 1264415 (1968).
961. Montecatini: Brit. Pat. 1015086 (1965).
962. Rudkowskii, D. M., Trifel, A. G., Gankin, W. J.: Brit. Pat. 1002691 (1965).
963. Takegami, Y., Watanabe, Y., Masada, H.: Bull. Chem. Soc. Japan **40**, 1459 (1967).
964. Cull, N. L., Mertzweiller, J. K., Tenney, H. M. (Esso): U. S. Pat. 3383426 (1968).
965. Pregaglia, G., Andreetta, A., Ferrari, G. (Montecatini): Chem. Commun. **1969**, 590.
966. Shell: Brit. Pat. 995459 (1965).
967. Heil, B., Marko, L.: Acta Chim. Acad. Sci. Hung. **55** (1), 107 (1968).
968. Esso: Brit. Pat. 1097364 (1968).
969. Falbe, J. (Ruhrchemie AG): Belg. Pat. 718857 (1969).
970. Nahum, L. S.: Org. Chem. **33**, 3601 (1968).
971. Stresinka, J., Marko, M., Macho, V.: Chem. Zvesti **22**, 263 (1968).
972. Houda, M., Koga, T., Noyori, G.: Kogyo Kagaku Zasshi **70**, 1346 (1967).
973. Kurokawa, H., Koga, T., Houda, M., Noyori, S.: Kogyo Kagaku Zasshi **70**, 1355 (1967).
974. Greene, C. R. (Shell): U. S. Pat. 3369050 (1968).
975. Day, J. P., Basolo, F., Pearson, R. G., Kangas, L. F., Henry, P. M.: J. Am. Chem. Soc. **90**, 1925 (1968).
976. Esso: Brit. Pat. 1072796 (1967).
977. Shell: Brit. Pat. 1002428 (1965).
978. van Winkle, J., Hasserodt, U. (Shell): German Pat. 1282633 (1968).
979. Booth, B. L., Else, M. J., Fields, R., Goldwhite, H., Haszeldine, R. N.: J. Organometal. Chem. **14**, 417 (1968).
980. Imyanitow, N. S., Rudkowskii, D. M.: Kinetika i Kataliz **8**, 1051 (1967).
981. Heil, B., Marko, L.: Magy. Kem. Lapja **1968**, 1252, 669.
982. Hieber, W.: Z. Anorg. Allg. Chem. **251**, 96 (1943).
983. Ullmanns Encyklopädie der technischen Chemie, Vol. 12, p. 320. München-Berlin: Urban & Schwarzenberg 1960.
984. Venanzi, L. M.: Chemistry in Great Britain, **4**, 162 (1968).
985. Wakamatsu, H., Sakamaki, K.: Chem. Commun. **1967**, 1140.
986. Day, J. P., Basolo, F., Pearson, R. G.: J. Am. Chem. Soc. **90**, 6927 (1968).
987. Ethylene and its Industrial Derivatives, p. 1169, edit. by S. A. Miller. London: Ernst Benn Ltd. 1969.
988. Fitzwilliam, J. W., Naragon, E. A., Moore, F. J. (Texas Co.): U. S. Pat. 2830089 (1958).

989. Alderson, T., Thomas, J. C. (Du Pont): U. S. Pat. 3040090 (1962), C. A. **57**, 16407 (1962).
990. Roos, L., Orchin, M.: J. Organometal. Chem. **31**, 3015 (1966).
991. Falbe, J. (Ruhrchemie AG): Belg. Pat. 718856 (1969).
992. Gankin, W. J., Genender, L. S., Rudkowskii, D. M., in: Carbonylation of Unsaturated Hydrocarbons, p. 61, edited by Allunions Sci. Red. Inst. for Petrochemical Processes. Leningrad: Chemistry Dept. 1968.
993. Möller, K. E.: Angew. Chem. **77**, 551 (1965).
994. Eidus, Y. T., Ordjam, M. B., Shokina, L. I., Kanevskaja, M. A.: Zh. Organ. Khim. **2**, 266 (1966).
995. Koch, H., Huisken, W., Möller, K. E., Lohbeck, K.: German Pat. 972315 (1959).
996. Schering AG: Fr. Pat. 1348894 (1963).
997. Swern, D.: Rev. Franc. Corps Gras. **8**, 7 (1961).
998. Koch, H., Schauerte, K.: Brennstoff-Chem. **46**, 392 (1965).
999. Pawlenko, S.: Chem. Ingr.-Tech. **40**, 52 (1968).
1000. — (Schering AG): German Pat. 1212061 (1966).
1001. Schauerte, K., Koch, H.: Brennstoff-Chem. **49**, 263—267, 302—305 (1968).
1002. Falbe, J.: Aliphatische Monocarbonsäuren. In: Ullmanns Encyklopädie der technischen Chemie, Ergänzungsband, 3rd edit., p. 120/130. München-Berlin-Wien: Urban & Schwarzenberg 1970.
1003. Hydrocarbon Process. **46**, 154 (1967).
1004. Timm, B.: Lecture to the 65th session of Deutsche Bunsengesellschaft für Physikalische Chemie (21-6-1966), referred in "Die BASF", no. 1, p. 23.
1005. Hydrocarbon Process. **46**, 140 (1967).
1006. Blick durch die Wirtschaft **259**, 5, (1968).
1007. Hohenschutz, H., v. Kutepow, N., Himmele, W.: Hydrocarbon Process. **11**, 141 (1966).
1008. Hydrocarbon Process. **46**, 136 (1967).
1009. Paulik, F. E., Roth, J. F.: Chem. Commun. **1968**, 1578.
1010. Schulz, R. G., Montgomery, P. D.: J. Catalysis **13**, 105 (1969).
1011. Chem. Age India, p. 20, Jan. 1969.
1012. Ellwood, P.: Chem. Eng. **76**, 148 (1969).
1013. Angew. Chem. **81**, 158 (1969).
1014. Chem. Eng. **72**, 78 (1965).
1015. Weber, H., Falbe, J.: Lecture given to the Amer. Chem. Soc. 158th National Meeting, New York, Sept. 7—12, 1969.
1016. Mori, Y., Tsuji, J.: Bull. Chem. Soc. Japan **42**, 777 (1969).
1017. Mizoroki, T., Nakayama, M.: Bull. Chem. Soc. Japan **41**, 1628 (1968).
1018. Imyanitow, N. S., Rudkowskii, D. M.: Zh. Prikl. Khim. **40**, 2825 (1967).
1019. Piacenti, F., Bianchi, M., Lazzaroni, R.: Chim. Ind. (Milan) **50**, 318 (1968).
1020. Chiusoli, G. P., Dubini, M., Ferraris, M., Guerrieri, G., Merzoni, S., Mondelli, G.: J. Chem. Soc. C/23, 2889 (1968).
1021. Bittler, K., v. Kutepow, N., Neubauer, D., Reis, H.: Angew. Chem. **80**, 352 (1968).
1022. Guerrieri, F., Chiusoli, G. P.: J. Organometal. Chem. **15**, 209 (1968).
1023. McClure, J. D.: J. Org. Chem. **32**, 3888 (1967).
1024. Matsuda, A.: Bull. Chem. Soc. Japan **42**, 571 (1969).
1025. Imyanitow, N. S., Rudkowskii, D. M., in: Carbonylation of Unsaturated Hydrocarbons, p. 198, 211. Edited by Allunions Sci. Res. Inst. for Petrochemical Processes. Leningrad: Chemistry Dept. 1968.

1026. Kuwajew, B. J., Imyanitow, N. S., Rudkowskii, D. M., in: Carbonylation of Unsaturated Hydrocarbons, p. 222, 225, 232. Edited by Allunions Sci. Res. Inst. for Petrochemical Processes. Leningrad: Chemistry Dept. 1968.

1027. Imyanitow, N. S., Jujajew, B. J., Rudkowskii, D. M., in: Carbonylation of Unsaturated Hydrocarbons, p. 215. Edited by Allunions Sci. Res. Inst. for Petrochemical Processes. Leningrad: Chemistry Dept. 1968.

1028. Fell, B., Seide, W., Asinger, F.: Tetrahedron Letters 8, 1003 (1968).

1029. Kunichika, S., Sakakibara, Y., Nakamura, T.: Bull. Chem. Soc. Japan 41, 390 (1968).

1030. The Petrochemical (Japan) Aug. 7, (1969).

1031. Oil, Paint, Drug Reptr. Febr. 3, (1969).

1032. VWD-Chemie 158, 4 (1969).

1033. Eur. Chem. News 391, 7 (1969).

1034. Fell, B., Rupilius, W.: Tetrahedron Letters 32, 2721 (1969).

1035. Tsutsui, M., Hancock, M., Ariyoshi, J., Levy, M. N.: Angew. Chem. 81, 435 (1969).

1036. Gregorio, G., Pregaglia, G., Ugo, R.: Inorganica Chimica Acta/3 : 1/, 89 (1969).

1037. Dümbgen, G., Neubauer, G.: Chem. Ingr.-Tech. 41, 974 (1969).

1038. Eur. Chem. News no. 400, p. 6 (1969).

1039. Chem. Eng. News 47, 24 (1969).

1040. Weiss, A. H.: Hydrocarbon Process. 48, no. 10, 125 (1969).

1041. Mistrik, E. J., Durmis, J.: Chem. Zvesti 23, 286—294 (1969).

1042. Palagyi, J., Marko, L.: J. Organometal. Chem. (Amsterdam) 17, 453—456 (1969).

1043. Ungvary, F., Marko, L.: J. Organometal. Chem. (Amsterdam) 20, 205 bis 209 (1969).

1044. Bianchi, M., Benedetti, E., Piacenti, F.: Chim. Ind. (Milan) 51, 613 (1969).

1045. v. Kutepow, N., Bittler, K., Neubauer, D. (BASF): U. S. Pat. 3 437 676 (1969).

1046. Falbe, J., Tummes, H., Hahn, D.: Angew. Chem. 82, 181 (1970); Angew. Chem. Intern. Ed. Engl. IX, 169 (1970).

1047. Chem. Eng. News 47, 20 (1969).

1048. Chem. Eng. 76, 174 (1969)

1049. Chem. Ind. Düsseldorf XXI, p. A 996 (1969).

1050. Chem. Eng. 76, 203 (1969).

1051. Piacenti, F.: Lecture given at Ruhrchemie AG, Oberhausen-Holten on 13-1-1970 and private communication.

1052. Falbe, J. (Ruhrchemie AG): German application, unpublished.

1053. Asinger, F., Fell, B., Rupilius, W.: Ind. Eng. Chem. Prod. Res. Develop. 8, no. 2, 214 (1969).

1054. U. S. Tariff Commission, Report on Synthetic Organic Chemicals, U. S. Production and Sales, 1967

1055. Hershman, A., Robinson, K. K., Craddock, J. H., Roth, F. J.: Ind. Eng. Chem. Prod. Res. Develop. 8, no. 4, 372 (1969).

Subject Index